HVAC and Energy Efficiency: A Practical Guide for Engineers and Facility Managers

Jaymin Pareshkumar Shah

DEDICATION

This book is dedicated to everyone who has played a role in shaping my journey—mentors, colleagues, family, and friends—whose unwavering support, guidance, and encouragement have been invaluable in my personal and professional growth.

To my family, who have stood by me through every challenge and success, your belief in me has been my greatest strength. Your love and sacrifices have fueled my passion for continuous learning and excellence.

To my mentors and colleagues, who have shared their wisdom, challenged my thinking, and inspired me to push the boundaries of energy management and sustainability. Your guidance has been instrumental in shaping my expertise and career path.

To all aspiring engineers, energy professionals, and sustainability advocates, may this book serve as a source of knowledge, inspiration, and motivation. The world of energy efficiency and sustainability is ever-evolving, and I hope to contribute to your journey in making a lasting impact on our environment and communities.

Lastly, to those who believe in the power of innovation and sustainability, let us continue striving for a greener, more efficient, and sustainable future—one project, one building, and one idea at a time.

Contents

PREFACE

Energy-efficient HVAC systems are needed now more than ever. Global energy demands are increasing, and sustainability is a must for engineering, creating a need for professionals in the building and facility sector to transform. A Practical Guide for Engineers and Facility Managers presents HVAC and Energy Efficiency as a practical, up-to-date resource filled with foundational HVAC knowledge, with modern strategies to optimize energy.

In years of observing these challenges first hand, this book was born. The chapters strive for theoretical and actionable techniques, best practices, and real-world case studies that can be put to immediate use in the reader's professional settings.

Atmospheric or large building energy supply systems are the focus of this guide for both practicing engineers as well as facility managers or students of the field or decision-makers developing building policies. It emphasizes smart systems, renewable integration, and technological advances that are going to make or break climate control in the built environment.

I believe that this book will function as both a reference and as a catalyst towards innovation, sustainability, and more efficient building systems for many industries.

Acknowledgment

The support, encouragement, and expertise of many individuals and institutions contributed to the possibility of writing this book.

Then, I owe a special thanks to the teachers and colleagues in the field of mechanical and energy engineering, whose knowledge and experiences helped to fill this work with content. I would like to thank the engineering pros and facility managers who opened up their case studies, data, and feedback during the development of this book.

I would also like to thank all the technical reviewers and editions who ensured that the manuscript is clear, precise, and relevant. We are currently shaping this book into a practical tool for today's HVAC and energy efficiency challenges with your contributions.

Thanks to my family and friends who have been patient, motivated, and have been believing in the project.

I am also grateful to the professional societies, academic institutions, and organizations whose work is dedicated to implementing sustainable building practices. Your continued work continues to show and lead the activation of change throughout the industry.

List of Abbreviations

Abbreviations	Full Form
HVAC	Heating, Ventilation, And Air Conditioning
VRF	Variable Refrigerant Flow
ERVs	Energy Recovery Ventilators
BMS	Building Management System
VAV	Variable Air Volume
AHU	Air Handling Units
CFM	Cubic Feet Per Minute
DDC	Direct Digital Control
AFUE	Annual Fuel Utilization Efficiency
VOCs	Volatile Organic Compounds
EER	Energy Efficiency Rating
SEER	Seasonal Energy Efficiency Ratio
RES	Renewable Energy Sources
EMS	Energy Management Systems
BEMS	Building Energy Management System
BAS	Building Automation System
ML	Machine Learning
ACH	Air Changes Per Hour
TRNSYS	Transient System Simulation Program
AFUE	Annual Fuel Utilization
VSDs	Variable Speed Drives
IAQ	Indoor Air Quality
HRV	Heat Recovery Ventilators
HEPA	High Efficiency Particulate Air Filter
DCV	Demand-Controlled Ventilation

R value	Thermal Resistance
ECM	Electrically Commutated Motors
IOT	Internet of Things
GHG	Greenhouse Gas
PV	Photovoltaic
F gas	Fluorinated Gases
Y friendly	User-Friendly
CFCs	Chloro Fluoro Carbons
HCFC	Hydrochlorofluorocarbon
HFCs	Hydrofluorocarbons
HFOs	Hydro fluoro olefins
GWP	Global Warming Potential
GHG	Greenhouse Gases
PCM	Phase Change Materials
PDM	Predictive Maintenance

CHAPTER 01

INTRODUCTION

1.1 Chapter Overview

The scope of this chapter is on the basic technologies and systems involved with maintaining thermal comfort, air quality, and energy use efficiency performance in buildings. The study starts with an exposition of different heating technologies, including heat pumps, furnaces, boilers, and radiant heating, and describes their operations, efficiencies, and appropriateness for the respective building types and climates. Then the discussion goes on regarding cooling technologies, especially air conditioning systems, chillers, and cooling towers, and mentions their energy demand and the efficiency of indoor temperature control. Ventilation and air distribution systems, as well, are addressed in the chapter, the role of which is to ensure the quality of the indoor air and to provide efficient temperature regulation. It also stresses the significance of good ventilation, good airflow, and good air distribution to create a favorable environment for both health and comfort. In addition, the chapter addresses energy consumption in buildings, in particular, HVAC systems that represent a primary reason for excessive energy consumption. As a growing trend, the integration of renewable energy sources (RES), collectively referred to as solar power, as part of the HVAC operations is also discussed to reduce energy consumption and attain environmental goals. The chapter concludes with the role of "energy management systems" (EMS) and "building energy management systems" (BEMS) in energy use optimization, building efficiency improvement, and the support of sustainability goals. The chapter is achieved through this comprehensive exploration that emphasizes upon capability of integrating efficient technologies and smart energy management practices to make sustainable and efficient building environments.

1.2   Introduction to HVAC and Energy Efficiency

Heating, Ventilation, and Air Conditioning (HVAC) systems are essential to keeping a comfortable, well-air-conditioned house. Temp, humidity, and air quality of residential, commercial, and industrial spaces are regulated with these systems and are vital parts of our everyday life. HVAC systems range from simple individual units, like window air conditioners, to complex centralized systems that serve entire buildings. HVAC systems are important, but they are also some of the largest buyers of energy in buildings. Heating and cooling are a significant portion of total energy use globally, and in residential buildings, can be up to 40–50% of total energy use.

HVAC systems find increased application in a wide variety of building types, including commercial, institutional, residential, and industrial structures. According to ASHRAE, (2020)The primary function of a building's HVAC system is to ensure that those within can maintain a comfortable temperature by modifying the outside air to match the needs of the building's occupants. A system that draws in air from outside, heats or cools it, distributes it to occupied places, and then either returns it to the outside air or uses it again, depending on the weather outside. Some factors that should be considered when choosing an HVAC system for a building include the local climate, the building's age, the owner's and designer's personal preferences, the available budget, and the structure's architectural style. (ASHRAE, 2020).

There are several ways to categorise HVAC systems based on the processes and distribution methods that are essential. The necessary steps involve heating, cooling, and ventilating. Additional procedures, including humidification and dehumidification, can be incorporated. This can be accomplished with the help of appropriate heating, ventilation, air conditioning, and dehumidification systems. When it comes to HVAC systems, the distribution system is crucial for delivering the right amount of air under the right circumstances. Variations in the kind of refrigerant and the means of delivery, including fan coils, air ducts, water pipelines, and air handling equipment, primarily determine the distribution system.

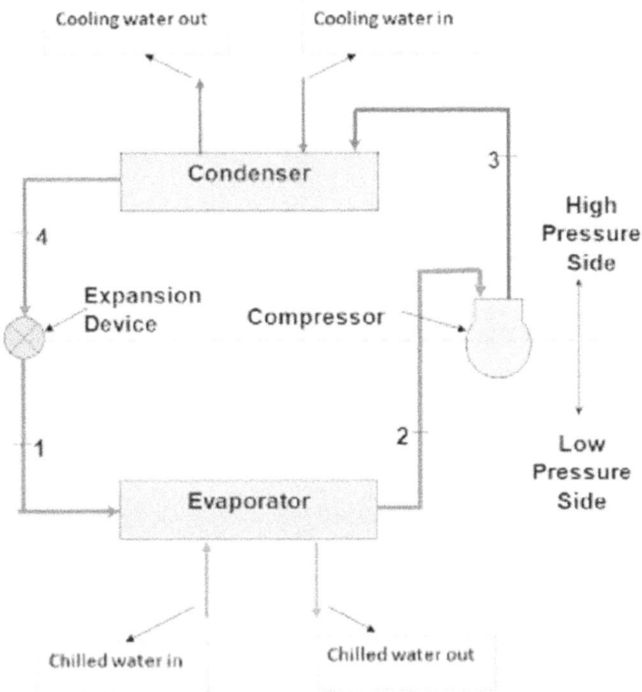

Figure 1.1: Energy Efficiency HVAC

Source: (Patel & Buddhi, 2022)

Overall, the energy efficiency of HVAC is a comprehensive process that requires efficient equipment, the use of smart technologies, better building insulation, and integration of renewable energy. The ongoing technological developments for HVAC equipment are leading to systems that consume less energy, thus helping businesses and residential sectors decrease expenses and environmental impact. With the increasing significance of building energy efficiency on the global scene, HVAC systems will play a pivotal part in reducing waste of energy and optimizing sustainability.

1.1.1 The Growing Importance of Energy Efficiency in HVAC

This interest in energy efficiency in HVAC systems is a symptom of the wider pressure to address climate change and energy conservation, and reduce costs. Energy consumption is increasing both globally, and HVAC systems that regulate the indoor environment have become a target location for energy optimization. An HVAC system is a large energy user in both residential and commercial buildings; estimates suggest that heating and cooling account for around 40-50% of energy use in homes. Therefore, HVAC energy efficiency improvement is a key strategy to save operational costs, minimize carbon emissions, and enhance sustainability.

Several factors push for the development of energy-efficient HVAC solutions, for example, rising energy prices, stricter environmental regulations, and growing adoption of green building practices. The use of energy-efficient HVAC systems reduces the demand for nonrenewable sources of energy, thereby decreasing greenhouse gas emissions and contributing to the mitigation of climate change. In today's context, with a global transition towards sustainable development, building owners and developers are looking more strongly to ramp up the use of energy-efficient technologies that synergize with environmental goals and save on long-term costs.

The financial payback of taking care of energy efficiency in HVAC is certainly one of the strongest reasons that made this factor so important. Businesses and homeowners can expect to save a lot by improving system performance and reducing energy consumption on utility bills. Thus, investing in energy-efficient HVAC technologies becomes a viable option when these costs represent a significant part of overall operational expenses in commercial settings where energy costs can be a major expense component. As a result, this has led to high-efficiency equipment like variable refrigerant flow (VRF) systems and energy-efficient heat pumps that can control temperature and airflow better and consume relatively less amount of energy for heating and cooling.

Plus, technological advancements have also played a big part in raising HVAC energy efficiency. The effect of smart technologies like smart thermostats and a system that is fully automated has completely changed or moved HVAC systems as to how the systems are controlled and how they are handled. These systems have real-time data along with sensors that adjust the temperature settings based on factors such as several people in a room, weather conditions, and even the time of day. For example, smart thermostats can learn user preferences and optimize energy use accordingly. In addition, BMS and IoT-enabled devices facilitate real-time monitoring and diagnostics, enabling one to identify inefficiencies and achieve the best possible performance.

HVAC energy efficiency also has a large effect on a macro scale, incorporating a renewable energy supply. Building owners are also able to make use of renewable technologies like solar panels and geothermal heat pumps as their costs lower. Combining the use of HVAC with renewable energy can prevent a high dependence on grid power to use conventional electricity generation and, therefore, reduce the carbon footprint and bring energy independence. Solar-powered HVAC systems are an example of such systems, utilizing the sunlight to power cooling systems, as geothermal heat pumps effectively use the Earth's inherent heat for both heating and cooling. (henick, 2023).

1.2.2 Role of Engineers and Facility Managers in Energy Conservation

The ongoing efforts of conserving energy in the buildings and the facilities are heavily dependent on the efforts of those engineers and facility managers in seeking to optimize the performance of the HVAC systems. Expertise and diligent management are key to the blue-sky development of energy-saving opportunities, adoption of sustainable practices, and the facilitation of energy-efficient technologies. These professionals are meeting this important need by either creating or maintaining energy-efficient environments, as more and more energy conservation is needed to reduce operational costs and meet environmental goals.(Goulden & Spence, 2015).

1. Role of Engineers in Energy Conservation

Most engineers who need to work with HVAC systems and other infrastructure (mechanical, electrical, and energy engineers) are required to design, install, and optimize those systems. It is important that they play a key role in ensuring energy-efficient technologies are used in all new buildings and retrofitted into existing buildings. What follows are the tips on how engineers are helping conserve energy:

Engineers are tasked with creating energy-efficient HVAC systems that choose the most appropriate equipment, like high-efficiency heat pumps, VRF systems, and energy recovery ventilators (ERVs), once they are designed. Building layout and insulation are also taken into consideration for their energy loss. With the help of advanced simulation tools, engineers solve for the energy consumption and system parameters that optimize the system for efficiency.

- **Middle Management:** Engineers implement energy-efficient technologies by evaluating and integrating cutting-edge technologies into smart thermostats, advanced building automation systems (BMS) and energy monitoring tools. These technologies allow the HVAC systems to be controlled precisely, maximizing energy consumption while keeping occupant comfort at a minimum.
- **Engineers perform** energy audits to assess how energy is being utilized in an entire facility. Engineers analyze data coming out of HVAC (and another energy-devouring) equipment and search for ways to correct inefficiencies and propose efficiencies in things like upgrading to more efficient equipment, making system adjustments, or changing the layout of the building to minimize the energy requirements.

- **Renewable Energy Routing:** In addition, Engineers design renewable energy integration, like solar thermal systems for water heating or geothermal heat pumps for heating and cooling. These solutions cut down on the reliance on energy from traditional sources while reducing the effects the building has on the environment.

2. Role of Facility Managers in Energy Conservation

The day-to-day operation, maintenance, and management of HVAC systems and other building utilities are the responsibility of facility managers. The energy conservation role involves continuous building performance optimization and ensuring the systems are already at maximum efficiency. Generally, however, facility managers are more hands-on and deal with the operation of the building and, therefore, are directly responsible for its energy usage. Key contributions include:

1. **Maintenance:** Facility managers monitor HVAC systems to maintain them, as well as to prevent issues before they arise. Preventative maintenance on the equipment, like cleaning and resetting filters, checking and replacing refrigerant, and inspecting gas ductwork, can help prevent energy waste and prolong the life of the equipment. An efficiently running well well-looked-after system entails the consumption of less energy than is required.

2. **Implementation of Energy-Efficient Technologies:** Facility managers make decisions in real time with the help of real-time building management systems (BMS) or other monitoring tools to operate by energy usage requirements. An example of this is that they can program temperature schedules according to occupancy patterns or set the temperatures so as not to overcook or overheat. Significant energy savings can be achieved without impacting occupant comfort.

3. **Energy Audits and System Evaluation:** Facility managers assist in employee and Tenant education to foster energy conservation

in a building by educating the employees and tenants on how to properly use energy. Collectively, various simple actions such as turning off lights when they are not in use, regulating thermostats at the most optimal levels, and using natural light to save energy can save loads of energy.

4. **Energy Performance Report:** This tracks and reports on the energy consumption of facility managers and compares it against benchmarks or historical data to find out trends and areas of improvement. Tracking these reports is key in determining whether or not the energy conservation efforts are working and if there is a need to upgrade or improve in the future.

5. **Involvement with Facility Savings/Health Initiatives:** Facility managers are regularly engaged in launching facility savings/ health initiatives, which involve installing energy-efficient lighting, renewable energy systems, water-saving technologies, etc. Engineers spend a lot of time communicating with them to make sure their energy-saving projects fit as a whole with how much energy a building should be using, while also ensuring that they are maintained properly as time goes on(Goulden & Spence, 2015).

Building energy conservation depends on the combined efforts of engineers together with facility management personnel. The technology design and implementation work of engineers enables energy efficiency, while facility managers' duties focus on the daily operational effectiveness of these systems. Through their combined actions, these two professions serve as vital elements that help decrease building power use and operational expenses and advance sustainability targets for building structures.

1.2.3   Key Challenges in HVAC Energy Optimization

HVAC systems require energy use to be optimized to reduce energy consumption, cut operational costs, and make it more environmentally sustainable. Nevertheless, some things prevent HVAC systems from realizing optimal energy efficiency. Typically, these barriers are technical, financial, operational, and regulatory factors; hence, limiting any ability to optimize energy across the board is a complex and time-intensive process. Some of the main challenges with HVAC energy optimization are given below:

1. High Initial Investment Costs

The high initial investment associated with such energy-efficient technologies is one of the biggest barriers to properly optimizing HVAC systems. It also may take considerable capital expenditure to upgrade to high-efficiency equipment like advanced heat pumps and variable refrigerant flow (VRF) systems or to install building automation systems (BMS). However, these costs can inhibit the business or building owners of limited budgets from adopting the energy-saving technologies, even despite potential long-term savings. Energy-efficient HVAC systems do not produce a return on investment (ROI) quickly, but the majority of organizations face the problem of incurring the initial cost.

2. Complexity of Retrofitting Existing Systems

HVAC systems in many buildings, especially older ones, were not intended from an energy-efficient point of view. These systems present a complex and costly challenge of retrofitting to improve energy performance. Even with the integration of modern technologies, including smart thermostats, sensors, and energy recovery ventilators (ERVs), the outdated systems require such extensive modification of the infrastructure that it disturbs the operations and adds to the costs. Because retrofitting is complex, system optimization may not ideally reduce the energy used without complete system alteration.

3. Balancing Energy Efficiency with Comfort

Consequently, the design of energy-efficient HVAC systems always has two goals: keeping people comfortable and reducing energy use.

This can be very difficult to strike a balance between energy savings and provided comfort levels (temperature, humidity, air quality, etc). High energy use is achieved through over-optimization, which is the practice of setting unrealistically low temperature limits or restricting the air exchange to minimize energy use. This is a delicate balancing act of energy optimization strategies that building owners and facility managers are obligated to ensure energy optimization strategies do not sacrifice the well-being of the occupants.

4. Lack of Real-Time Monitoring and Data Integration

If this data is accurate and real-time and accurate, it enables effective energy optimization. In the absence of good monitoring tools and integration, it is very difficult to identify inefficiencies, track energy consumption patterns, and make the necessary adjustments in time. BMSs and IoT-based sensors flaunt data, and yet there's no integration between these to get an overall view of energy use. Additionally, facility managers may not be able to notice small inefficiencies that, cumulatively, cause huge waste of energy, without continuous data analysis.

5. System Complexity and Technological Integration

For many modern HVAC systems, there are many components involved: variable speed drives, heat recovery systems, advanced controls, as well as sensors. The facility manager managing these complex systems needs the right specialized knowledge and technical expertise, which usually do not come easily for a building manager who has not been trained fully in energy management or system optimization. Moreover, it can be a challenge in terms of technology when new technologies are combined with other existing systems, especially when such legacy equipment is not compatible with newer, energy-efficient scales.

6. Regulatory and Compliance Challenges

HVAC energy optimization is usually based on a large number of local, national, and international regulations, codes, and standards. In particular, such regulations can be difficult to comply with for building owners and facility managers; particularly, the specific energy efficiency requirements can depend on the area, type of building, and its age-to-build system. Furthermore, if regulations start to get tighter, then buildings might need to be remodeled to meet the most up-to-the-minute energy standards, which can be a great deal of investment and preparation. It is also no wonder that there is often a risk of noncompliance with energy efficiency regulations, given the changing landscape of energy efficiency regulations.

7. Short-Term Focus on Cost Savings

Most organizations give much attention to short-term financial issues instead of long-term energy savings. This short-term focus can prevent the adoption of energy-efficient HVAC systems since the energy savings realized from the energy optimization will not be apparent immediately. For example, businesses may sacrifice long-term financial benefits for energy efficiency for short-term operational savings, postponing investments that are needed. Such an attitude can lead to a scarcity of support for energy optimization initiatives, especially when stakeholders do not perceive the potential for their long-term cost savings or are reluctant to spend the money to achieve such projects.

8. Inconsistent Occupancy and Usage Patterns

Due to a mismatch between the peak energy demand that a building experiences and the design of the HVAC system, which is optimally chosen to meet it, processes can be less efficient when occupancy levels change. Office buildings, schools, and hotels are examples of variables where occupancy changes with time, and HVAC system parts continue to run at full capacity of operation during low occupancy periods. In these situations, more energy will be used for the system to achieve a comfortable environment. Real-time controls that enable fine-tuning of HVAC systems to adjust to occupancy patterns, weather conditions, and usage schedules are currently not an easy problem.

9. Maintenance and System Degradation

Just as any mechanical system, if ever maintained, it undergoes wear and tear, and over time it tends to become less and less efficient. Regular maintenance is necessary to discharge the components of filters, coils, and compressors to ensure energy efficiency. To cut down the energy waste, facility managers must keep systems in good order and fix them as soon as possible. But maintenance of HVAC will be time-consuming, resource-intensive, and a time-consuming process that requires expertise and money. Therefore, schedules for regular maintenance must be adhered to in terms of ensuring that energy optimization strategies are effective and not short-lived, where they lose out over time.

10. Behavioral Barriers

HVAC systems are difficult to bring energy efficiency to optimization due to human behavior. By leaving windows open, by adjusting the thermostats to extend beyond recommended levels, or by operating space heaters and fans simultaneously with HVAC systems, occupants may unconsciously increase inefficiency. The importance of practical energy-saving behaviors is important to optimizing energy consumption and must be taught to the building occupants. Yet, implementing energy efficiency measures such as improvements to HVAC comes with tough work, and the difficulty in changing behavior and low levels of engagement from building occupants can undermine these efforts(Ian Dempster, 2018).

Technical, financial, operational and regulatory complexity make HVAC energy optimization challenges, but opportunities for improvement can be realized when tackled systematically. There are government incentives, energy saving grants, and ROI minded financial models that help to overcome high initial cost. Modular solutions and phased upgrades of existing systems are required for retrofitting that is complex and disruptive in order to minimize costs and disruptions. With smart technologies such as adaptive thermostats and the IoT enabled monitoring, we can achieve a balanced energy efficiency and occupant comfort in this way, which is based on the environmental as well as occupancy factors.

This, moreover, leaves gaps in real time data integration which can be closed by installing more advanced BMS and predictive analytics with the help of AI. Thus, training and certification programs for engineers and facility managers should focus on how to manage modern HVAC systems well. Proactive planning, that is, with alignment to industry standards, can also streamline regulations and compliance.

Finally, through education and incentives to occupants, as well as robust maintenance schedules, the energy efficiency is sustained. In the presence of these integrated approaches, the existing barriers will not be overcome but also in line with the broad objectives of sustainability, cost reduction, and operational excellence in the HVAC energy optimization.

1.3   Basics of HVAC Systems

The field of mechanical engineering known as HVAC is responsible for creating a comfortable indoor climate for building occupants by the systematic regulation of air movement, temperature, humidity, and air quality. (Aireserv, 2024).

The following is a list of the most essential parts of an HVAC system, which provide conditioned air to meet the thermal comfort needs of a building's occupants and to ensure good air quality within the building:

a.   Mixed-air plenum and outdoor air control
b.   Air filter
c.   Supply fan
d.   Exhaust or relief fans and an air outlet
e.   Outdoor air intake Ducts
f.   Terminal devices
g.   Return air system

> h. Heating and cooling coils
> i. Self-contained heating or cooling unit
> j. Cooling tower
> k. Boiler m. Control
> l. Water chiller
> m. Humidification and dehumidification equipment

The field of mechanical engineering and MEP encompasses it. All heating, ventilation, and air conditioning systems have these three primary parts.

- **Heating:** Heating is provided by a central heating system, and ventilation is the process of mechanically circulating air.
- **Ventilation:** The primary function of mechanical ventilation systems is to circulate air in a structure in a systematic manner. Using mechanical ventilation systems, air may be exchanged to keep the building's occupants in a comfortable and pleasant environment.
- **Air conditioning:** For MEP and HVAC systems, it is a must-have component for year-round temperature control, but particularly in the warmer months. Indoor air quality is improved by air conditioning systems, which also cool the area, remove humidity, and offer ventilation.

➤ Energy Efficiency in HVAC Systems

Reducing energy consumption and, in turn, energy costs and environmental effects can be achieved through HVAC system energy efficiency. HVAC systems account for a significant amount of energy consumption in residential, commercial, and industrial buildings. Reducing greenhouse gas emissions and, in turn, reducing their financial savings are the benefits of enhancing their efficiency. The performance of HVAC systems can be optimized through energy-conserving strategies, and several factors influence the efficiency of the HVAC systems(Connor Holbert, 2025).

➢ HVAC system requirements

There are four necessities for every HVAC system. Figure 1.2 depicts the essential machinery, space needs, air distribution, and pipe requirements.

Heating equipment, including steam and hot water boilers, is considered primary equipment. Air delivery equipment, including centrifugal, axial, plug, and plenum fans, and refrigeration equipment, is used to cool and condition the air that is delivered into space. It contains refrigeration-process refrigerants and water-based cooling coils from water chillers. The size of the available space is a major consideration when designing a central or local HVAC system. It necessitates the following five facilities:

a. **Equipment rooms:** considering that the overall space needed for mechanical and electrical systems can be anything from 4% to 9% of the entire building footprint. To simplify shaft plans, centralize operation and maintenance, and shorten the length and size of duct, pipe, and conduit lines, it is best to house it in the middle of the building.

b. **HVAC facilities:** For heating and cooling a building, many facilities are necessary for refrigeration and heating equipment. Water chillers or cooling water towers for big buildings, condenser water pumps, heat exchangers, air-conditioning equipment, control air compressors, and other miscellaneous equipment are needed for refrigeration, whereas boiler units, pumps, heat exchangers, pressure-reducing devices, and other miscellaneous equipment are needed for heating. When planning the layout of the rooms that will house these pieces of machinery, it is important to keep in mind their respective sizes and weights, as well as any installation or maintenance requirements, as well as any relevant rules for combustion and ventilation air requirements.

c. **The HVAC** fan and associated auxiliary machinery are housed in the fan room. The rooms need to think about how big the fan shafts and coils are, how to install and remove them, and how to replace and maintain them. The airflow rate needed to condition the building determines the size of the fans, which can be either centrally or locally installed, depending on factors including cost, location, and availability. The convenience of being able to step outside is highly valued.

d. **Vertical shaft:** facilitates the installation of water and steam pipe systems and the distribution of air. Heating, ventilation, and air conditioning (HVAC) ductwork includes supply, exhaust, and return air ducts. Distribution via pipes comprises steam supply, condenser return, condenser water, chilled water, and hot water. In addition to the mechanical and electrical distribution that serves the entire building, the vertical shaft also contains the pipes for plumbing, fire prevention, and electric conduits and closets.

e. **Equipment access:** For installation, replacement, and maintenance purposes, the equipment room must have enough space to move big, heavy equipment.

One aspect of air distribution is the ductwork that carries the conditioned air to the specified location in the most efficient, least noisy, and least expensive method feasible. In air distribution, various terminal units can be used to deliver air at different velocities. Some of these units include grilles and diffusers, which allow for low-velocity supply air delivery; fan-powered terminal units, which use an integral fan to guarantee supply air; variable air volume terminal units, which allow for variable amounts of air delivery; all-air induction terminal units, which control primary air, induce return air, and distribute mixed air into space; and air-water induction terminal units, which contain a coil in the induction air stream. Insulating the building's ductwork and pipes will stop heat loss and reduce energy costs. An additional consideration is the availability of adequate ceiling spaces to accommodate ducting in both the suspended ceiling and the floor slab. These spaces can also double as return air plenums, so minimising the need for return ductwork.

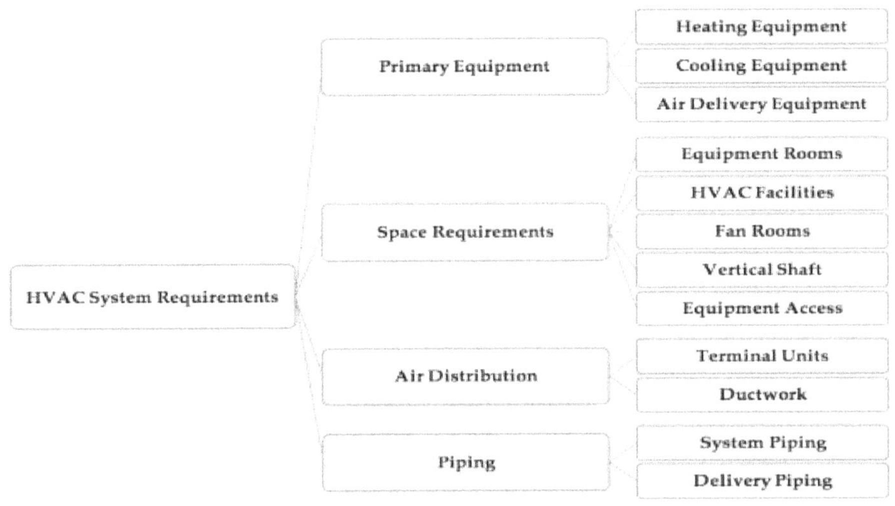

Figure 1.2: HVAC system requirements.

Source: (Seyam, 2018)

A direct, silent, and cost-effective method of delivering steam, condensate, hot water, cooled water, gas, and refrigerant to and from HVAC equipment is the piping system. The piping in the plant's main equipment room and the piping that leads to the delivery systems are the two primary types of pipes. Depending on the current regulations, HVAC pipe might or might not need insulation.

Benefits of Energy-Efficient HVAC Systems

There are plenty of benefits to adopting solutions with energy-efficient HVAC. This means that you will also see lower energy bills in terms of reduced electricity and fuel consumption, and that will mean large cost savings over time. Environmental benefits are: Less Carbon Footprint and fewer greenhouse gas emissions. Moreover, energy-efficient systems are more long-lived, which means less wear and tear, which in turn helps decrease maintenance costs. Improved indoor air quality and improved occupant comfort are direct results of optimization of the HVAC system.

Additionally, energy-efficient HVAC systems are generally installed in buildings, which allows them to have higher property values and better market appeal. Implementing sustainable HVAC solutions by businesses not only helps their businesses and oil and gas companies reduce energy costs, but it also benefits corporate reputation and meets growing energy regulations.

1.3.1   Types of HVAC Systems: Centralized vs. Decentralized

Various HVAC systems are available, each tailored to address the unique HVAC demands of building size, local temperature, and energy efficiency standards. It is possible to achieve peak efficiency, comfort, and performance with the correct HVAC system. The following are examples of HVAC systems:

1. Central HVAC Systems

Large buildings, commercial spaces, or even multiple-room homes that require creating a central heating and cooling system tend to use the central heating and ventilating of different air conditioning systems. Commonly, these are systems that feature a network of ducts for distributing conditioned air through different zones of the building to achieve equality of indoor climate control. They can offer High energy efficiency, better indoor air quality, and better temperature regulation than individual room units.

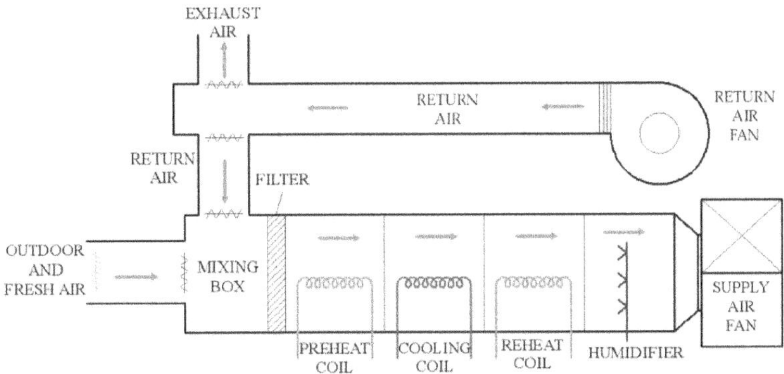

Figure 1.3: Central HVAC Systems

Sources: (Seyam, 2018)

a. **Split System:** In a split system, the interior unit and the outdoor unit are together. Both the interior and outdoor units typically contain coils for evaporation and condensation, with the former housing the former and the latter housing the air handler. The furnaces or heat pumps in these systems deal with heating as well as cooling, and these systems do a great job of keeping the temperature regulated. Among the most used HVAC configurations is the split system, as it is energy efficient, quiet, and can be zoned. Also, many modern split systems come with programmable thermostats and smart technology so that the user can control the temperature from the comfort of their home using mobile apps.

b. **Packaged HVAC System:** A packaged HVAC system comprises an enclosure with all major parts of an HVAC system, which are packaged into a single outdoor unit. For properties with limited indoor space, these units normally get mounted on the roof or the outside of the building. The packaged systems vary in configuration: straight cooling, gas-electric hybrid, and heat pump, and the systems are made to suit the various climatic conditions. A common use of them occurs in commercial buildings, schools, and retail spaces where centralized cooling is required, but not allowing for split systems in the available space is a predicament. The major advantages of packaged systems are ease of installation, lower cost of maintenance, and lower operation cost.

2. Ductless HVAC Systems

They are also known as ductless HVAC systems or mini split systems, which means there are heating and cooling options without ductwork. Because these systems are highly energy efficient and very versatile, they make an excellent choice for older buildings, room additions, apartments, and places where conventional ducted systems are impractical.

Figure 1.4: Ductless HVAC Systems.

Source: (carrier, 2023)

a. **Ductless Mini-Split System:** In a mini-split system, refrigerant lines connect an outdoor compressor/condenser to an indoor air handler or air handler. Indoor units operate independently to offer zoned heating and cooling. This arrangement provides better energy efficiency so that users can warm or cool particular rooms as required instead of the whole building. Minor split systems are quiet to set up and give superior atmosphere control, making them the perfect choice for commercial and residential homes, and small business structures.

b. **Multi-Split System:** A mini-split is all the same as a multi-split, except that a multi-split has an outdoor unit that connects multiple indoor units. With such a design, though, it becomes more efficient than two air conditioners, one for each room. Advanced inverter technology enables multi-split systems to regulate the speed of the compressor to optimize energy consumption and reduce operational costs. In such scenarios, these systems fit well in hotels with multiple rooms, large office spaces where cooling needs are different in different zones, etc.

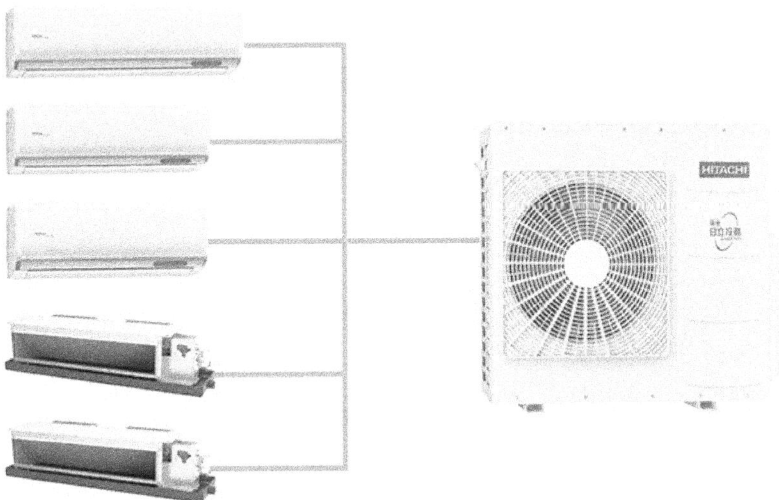

Figure 1.5: Multi-Split System.

Source: (Hitachi, 2023)

3. Hybrid HVAC Systems

Hybrid HVAC systems are combinations of different heating and cooling technologies to achieve maximum energy efficiency and saving. Thanks to these systems, the use of energy sources is automatically switched between different seasons.

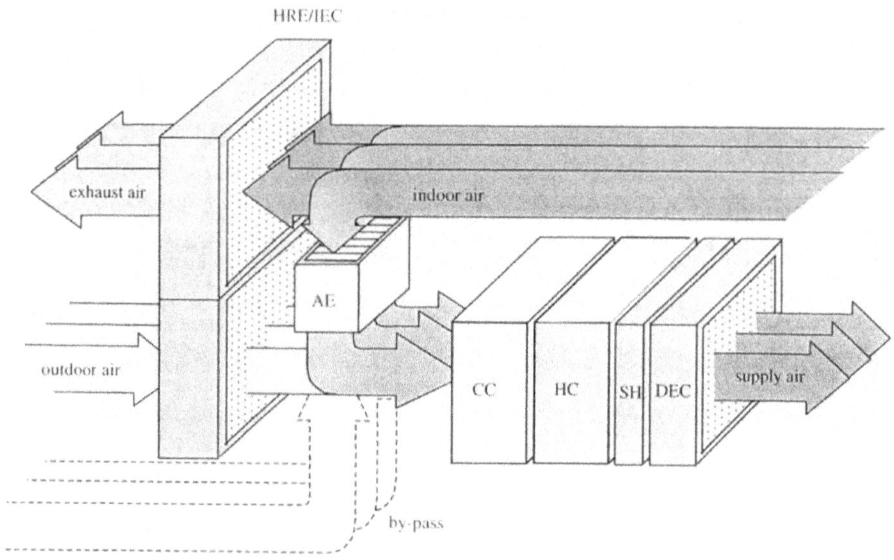

Figure 1.6: The hybrid HVAC system.

Source: (Palombo, 2002)

a. **Hybrid Split System:** A hybrid split system functions much the same as a normal split system, but it features a gas furnace in addition to a heat pump. It intelligently decides whether to use electricity or gas to heat according to temperature conditions for better energy efficiency and money saving. The beauty of this type of system lies in the fact that it delivers the best of indoor comfort in the sense it consistently keeps rooms comfortable, as well as making it possible for homeowners to save energy.

b. **Dual-Fuel System:** A dual-fuel HVAC (DUAL FUEL HVAC) is defined as a heat pump with a secondary heat source (typically a gas furnace or a boiler). In mild weather conditions, the system serves as a heat pump to transfer electricity as heat and cooling. In very cold temperatures, the system will switch to the second side heat source to provide further heat, even though the heat pump has become less so, much less efficient. As part of this setup up you can ensure maximum energy efficiency and save cost while being assured with a promise of proper warmth during winter months.

4. Geothermal HVAC Systems

The earth's natural underground temperature is used in geothermal HVAC systems for efficient and sustainable heating and cooling. These systems are some of the most environmentally friendly HVAC options in terms of long-term energy savings and they more than cut down on a system's carbon footprint.

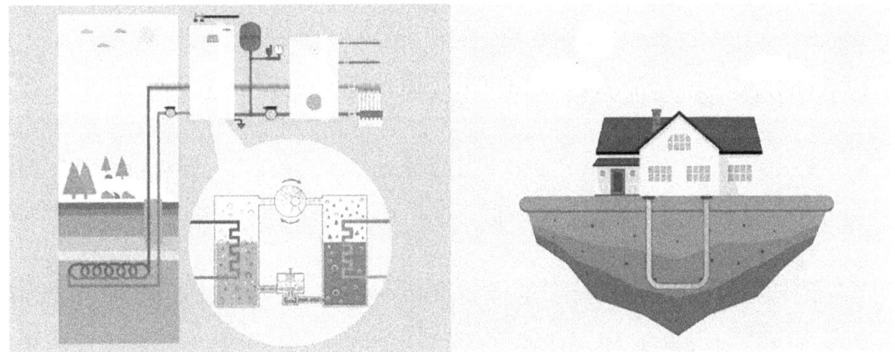

Figure 1.7: Geothermal HVAC Systems

Source: (RSI, 2024)

a. **Geothermal Heat Pump System:** A closed-loop or open-loop piping system transfers the heat between the building and the ground as it transfers heat between to geothermal heat pump system.

- In winter, the earth is extracted from the ground and used for indoor heating.
- In summer, it cools the space by sending in air, removing the heat from the building, and discharging it into the ground.

Because of its high initial installation costs (excavation and underground pipe installation), initial costs are high, however, geothermal systems have lower operating costs, higher efficiency, and long-term sustainability. Because they are well-suited to large residential properties and to eco-friendly commercial buildings and institutions wishing to eliminate energy dependence.

5. The systems that use variable refrigerant flow (VRF) and variable refrigerant volume (VRV)

VRF and VRV systems utilize advanced refrigerant flow technology to provide highly efficient and precise climate control. These systems are widely used in commercial buildings, hotels, hospitals, and office complexes due to their scalability and energy-saving capabilities.

Key Features of VRF/VRV Systems

- **Energy Efficiency:** By adjusting refrigerant flow based on demand, VRF systems optimize energy usage, reducing overall electricity costs.
- **Zoning Capabilities:** The ability to independently heat or cool different parts of a building allows for improved temperature management.
- **Heat Recovery VRF Systems:** These allow simultaneous heating and cooling in different zones, further enhancing energy efficiency and occupant comfort(Wan et al., 2020).

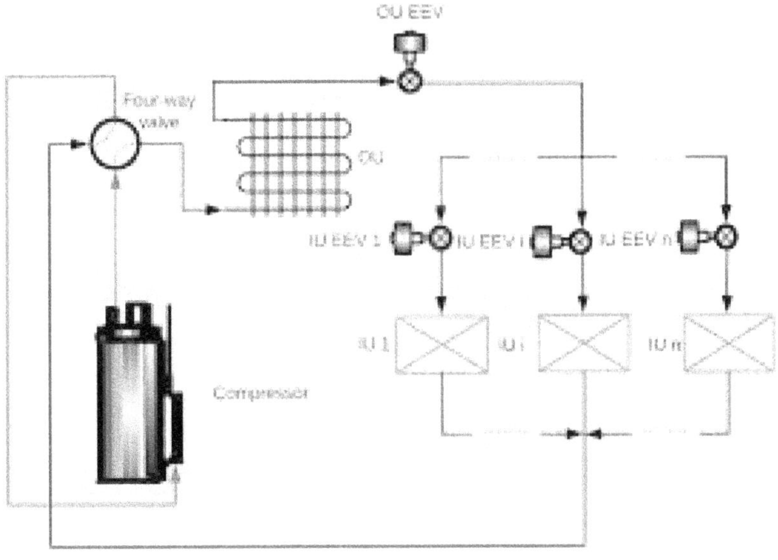

Figure 1.8: Variable Refrigerant Volume

Source: (Wan et al., 2020)

These systems provide **quiet operation, enhanced energy savings, and precise climate control**, making them an excellent choice for buildings with **complex climate control needs**.

Classification of HVAC systems

HVAC systems are primarily divided into two categories: central systems and decentralized or local systems. The main equipment location determines the type of system, which can be decentralized, conditioning a particular area of a building independently, or centralized, conditioning the entire structure as a whole unit. As a result, system classification and the location of key equipment should guide the design of the air and water distribution system. When choosing between two systems, the aforementioned factors ought to be used as well. The comparison of central and local systems based on the selection criteria is displayed in Table 1.1.

Table 1.1: Centralized vs. Decentralized

Criteria	Central System	Decentralized System
Temperature, Humidity, and Space Pressure Requirements	Fulfilling any or all of the design parameters	Fulfilling any or all of the design parameters
Capacity Requirements	Considering HVAC diversity factors to reduce the installed equipment capacity	Maximum capacity is required for each equipment.
	Significant first cost and operating cost	Equipment sizing diversity is limited.
Redundancy	Standby equipment is accommodated for troubleshooting and maintenance	No backup or standby equipment
Special Requirements	An equipment room is located outside the conditioned area, or adjacent to or remote from the building.	Possible that no equipment room is needed
	Installing secondary equipment for air and water distribution, which requires additional cost	Equipment may be located on the roof or ground adjacent to the building.
First Cost	Considering a longer equipment service life to compensate for the high capital cost	Affordable capital cost
Operating Cost	More significant energy-efficient primary equipment, A proposed operating system that saves operating cost	Less energy-efficient primary equipment, Various energy peaks due to occupants' preference, Higher operating cost

Maintenance Cost	Accessible to the equipment room for maintenance, keeping equipment in excellent condition, which saves maintenance costs	Accessible to equipment in the basement or living space. However, roof locations can be difficult due to bad weather.
Reliability	Central system equipment can be an attractive benefit when considering its long service life.	Reliable equipment, although the estimated service life may be less
Flexibility	Selecting standby equipment to provide an alternative source of HVAC or backup	Placed in numerous locations to be more flexible

Source: (Seyam, 2018)

1.3.2 Components of an HVAC System (Chillers, AHUs, VAVs, VRFs, etc.)

Indoor comfort, air quality, and climate control of a building are achieved with the help of an HVAC system. The system has components such as chillers, air handling units (AHUs), "variable air volume" (VAV) systems, and "variable refrigerant flow" (VRF) systems, all of which contribute to regulating temperature, humidity, and air distribution. All of these components combine so that the indoor environment continues to be comfortable and energy efficient, with healthy requirements. Cold air is brought into the facility with chillers and removed from the air by these chillers; these chillers remove the heat as well, while the AHUs deal with airflow and air quality by filtration as well as ventilation. VAV systems vary the amount of air supplied to different zones by demand, and VRF systems provide flexible, energy-efficient heating and cooling by adjusting refrigerant flow to those zones of the building.

This integration of these components should provide highly configurable HVAC systems that fit the particular needs of various buildings, such as residential, commercial, or industrial. Modern HVAC systems have advanced in terms of technology in that they are highly energy efficient in their components and thereby reduce environmental effects and costs of operation. One example of such equipment is VRF (variable refrigerant flow), which can precisely control heating and cooling to use less energy. Further invigorations of energy efficiency and indoor air quality include "energy recovery ventilators" (ERVs) and "demand-controlled ventilation" (DCV). Each of these components, when combined, also works to create an interconnected system both for comfort and sustains itself by decreasing energy consumption and increasing the overall performance of the system. The following are the key components commonly found in HVAC systems:

1. **Chillers:** The air is cooled in an HVAC system by chillers. How they work is that they remove heat from the air and pass it into a refrigerant, which is then looped through a cooling system to chill it down. In larger commercial buildings and large industrial applications, cooling systems that use many units, chillers, are used. Different types of chillers are available, such as air-cooled and water-cooled, with which applications are made suitable for building needs and climate.

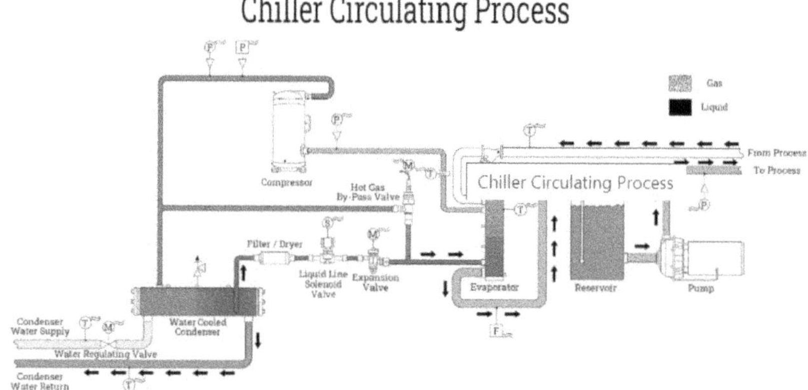

Figure 1.9: Chiller Circulating Process.

Source: (iqsdirectory, 2023)

Types of Chillers:

- **Air-Cooled Industrial Chillers:** When dissipation of heat is not an issue, an industrial chiller cooled by air is used. It does this by drawing heat from the water in the circulation and releasing it into the atmosphere. After the evaporator takes heat from the cooled water, the condenser releases it into the air when the refrigerant condenses.

Figure 1.10: Air Cooled Chiller

Source: (iqsdirectory, 2023)

- **Water-Cooled Industrial Chillers:** It is common practice to use a condenser water treatment system to remove mineral deposits from water-cooled industrial chillers, which are usually connected to a cooling tower. The water is chilled in the chiller once it has been supplied by the cooling tower.

Figure 1.11: Industrial Water Chiller

Source: (iqsdirectory, 2023)

- **Screw Industrial Chillers:** Screw industrial chillers, available in both water-cooled and air-cooled versions, utilize a helical rotor to compress and circulate refrigerant vapors.

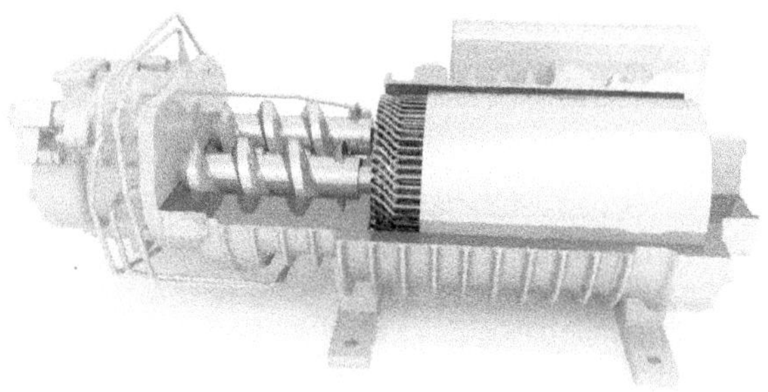

Figure 1.12: Screw Chillers

Source: (iqsdirectory, 2023)

2. **Air Handling Units (AHUs):** AHUs play an important role in circulating air within a building and conditioning it (heat or cool it). They help in the airflow regulation as well as the proper distribution of airflow with ducts from one corner of the building to another. The basic design of an AHU is a big metal container housing various parts that work together to accomplish a common goal.

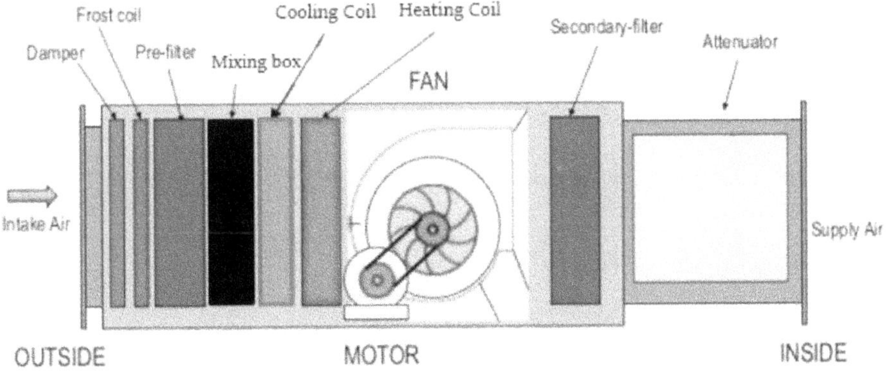

Figure 1.13: Air Handling Units (AHUs)

Source: (Paddeco, 2021)

This "metal box" is directly linked to the ducts that distribute conditioned air to designated areas of a building. Detailed descriptions of each of the main parts of an AHU follow the cross-sectional diagram that follows. In terms of AHU, two distinct varieties exist:

a. **Single Zone Systems and Multi Zone Systems:** A single duct connects the air handling unit to the entire structure in a single zone system. As it travels through the structure, the duct divides into several sections to supply each zone. Although it might be more convenient to use a single duct throughout the structure, doing so has its restrictions, like as ensuring that every room or zone stays at a constant temperature.

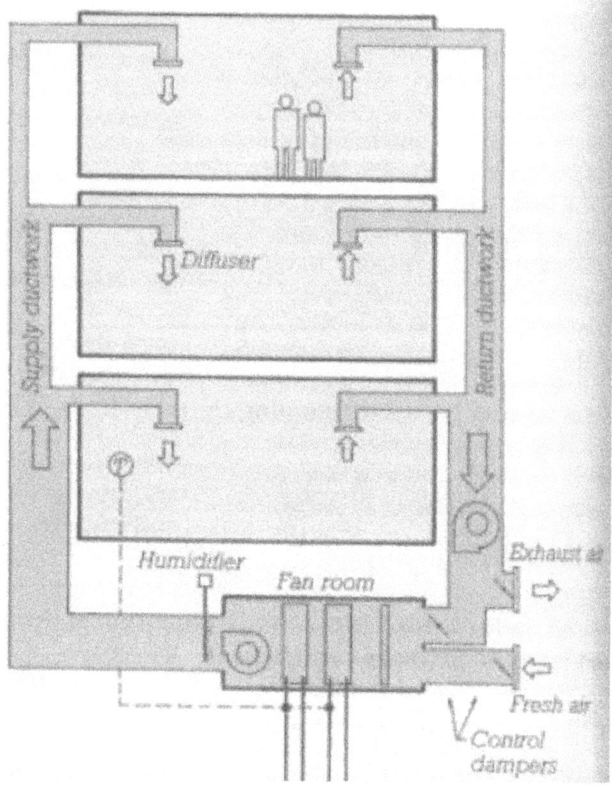

Figure 1.14: Single Zone Systems and Multi-Zone Systems

Source: (Paddeco, 2021)

b. **Multi-Zone System:** The AHU supplies each zone or room with its own set of ducts. While this system does increase the amount of labour needed on-site for installation, it offers the advantage of allowing for individual temperature control of each zone.

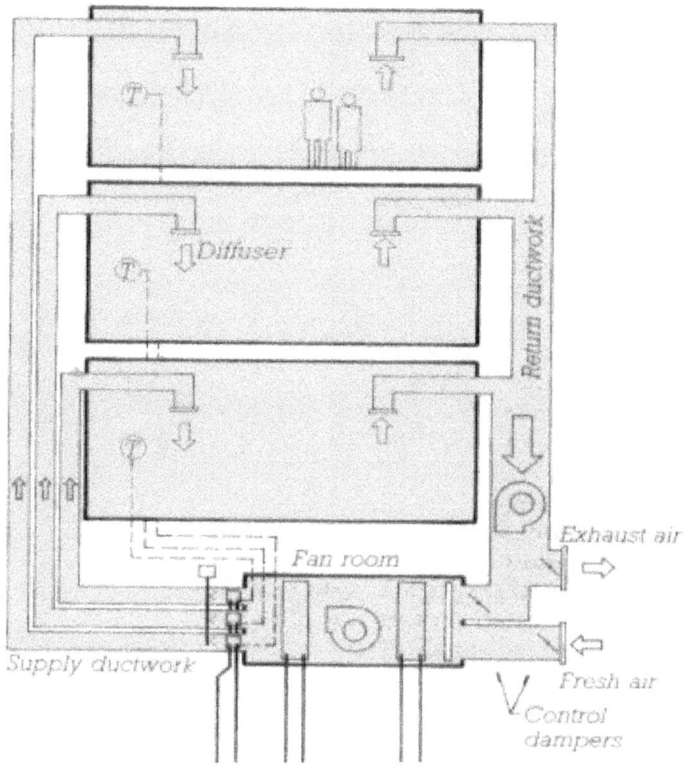

Figure 1.15: Multi-Zone System

Source: (Paddeco, 2021)

3. **Variable Air Volume (VAV) Systems:** The VAV system controls airflow in spaces by adjusting supply air volumes according to different heating and cooling requirements. The system uses a better method to manage thermal comfort and airflow in expansive structures through reduced energy usage.

Function and Components of a VAV Box

- **Airflow Sensor** – measures the airflow at the box's intake and uses that data to set the damper's location. The velocity pressure, which the controller uses to calculate CFM via the VAV box's intake, is determined by the airflow sensor, which measures both static and total pressure. Pressure at a certain velocity is equal to the sum of all pressures minus the amount of static pressure.

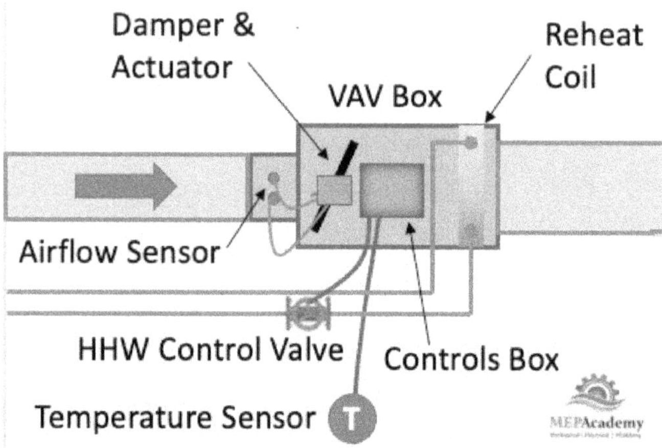

Figure 1.16: Variable Air Volume VAV Terminal

Source: (Instructor, 2022)

- **Actuator** – Based on the percentage of space demand and airflow, the damper will power a rotation.
- **Damper** – Air flow (CFM) controlled by the temperature sensor and the air flow sensor input.
- **Reheat Coil** – Depending on the zone, there may be a reheat coil that provides heating from heating hot water, steam or electricity. Some jurisdictions prohibit the use of electricity due to energy codes.
- **VAV Box Controller** – The controller will receive input (for a given temperature sensor and airflow sensor) and then send an output signal to the damper or could even send a signal to the heating hot water valve to modulate open or close. Controls can be pneumatic, electronic, or direct digital control (DDC). The older type of control is pneumatic, which is now in use but DDC system is at greater energy efficiency.

Fans and filters would constitute other components on a fan-powered box used in other versions of the VAV box.

4. **Variable Refrigerant Flow (VRF) Systems:** Similar concept to VRF systems with the main difference being that the refrigerant is supplied to an external condensing unit, then distributed to different plants or internal evaporators (usually Fan coil units). Each thermal zone of the building feeds a different internal evaporator, the refrigerant is flowed to each evaporator according to the local need. That's a lot of flexibility and enables systems at large to run as efficiently as possible by matching the overall internal demand with the output of the outdoor condenser. Second, VRF systems can generally be two-pipe or three-pipe systems, very broadly.

- There can be two types of pipe systems available for cooling or heating (heat pump), all of the zones.
- Three pipe systems can supply heating and cooling simultaneously or alternately heat some zones and cool others with heat recovery , which allows heat produced in cooling zones to be used to heat warming zones. Although more capital cost, heat recovery allows extremely efficient operation (low operating costs).

For buildings with several spaces, being of different thermal loads and requiring good local control, VRF systems are most suited to properties where some rooms can be unoccupied, whilst others have high thermal demand, for example hotels. They may also be suited to use in retrofitting older buildings due to their limited space requirements (depending on how ventilation is provided) vs some other systems.

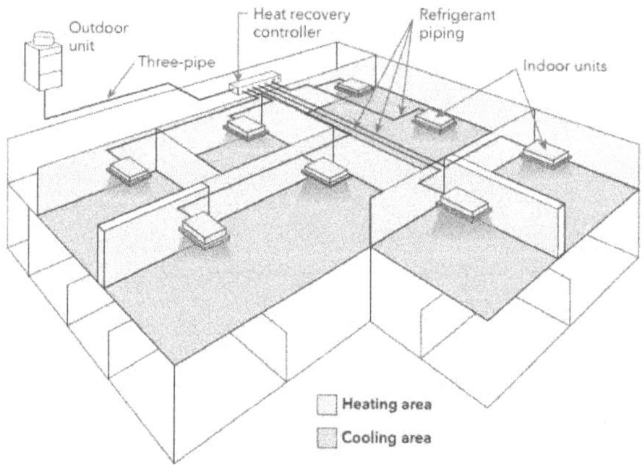

Figure 1.17: Variable Refrigerant Flow (VRF) Systems

Source: (slipstreaming, 2023)

5. **Other Key Components:** Beyond these main components of an HVAC system, there are other parts such as ducts for the distribution of air, thermostats for controlling the temperature, filters that enhance indoor air quality, humidifiers and dehumidifiers for controlling the humidity levels, and pumps for circulating the air or refrigerants. Building management systems (BMS) for automatic and data-based optimization of overall HVAC performance and energy recovery ventilators (ERVs) for efficient ventilation that also saves energy contribute to modern systems.

These are important components that must be present for the HVAC system to function properly in as they help the system operate efficiently and preserve a comfortable environment with a lower energy requirement and operating costs. However, the right component selection, integration, and maintenance of these components are necessary for optimum HVAC system performance.

1.3.3 Heating Technologies: Boilers, Heat Pumps, and Radiant Heating

During the colder months, it is essential to maintain comfortable indoor temperatures which can be done by means of heating technologies. The working principle is for generating or transferring heat to provide a warm and cosy environment inside buildings. There are different heating systems that are dependent on different energy sources such as electricity, natural gas, oil, solar energy, and geothermal energy. Which means that the right heating technology depends on the type of building, the local climate and energy efficiency goals.

Types of Heating Systems

It is recommended to have a variety of home heating solutions. There are distinctive qualities to each of them.

1. Heat Pumps

Heat pumps are an energy efficient means to transfer heat from one area to another rather than creating it. During winter, they heat the inside of their home with outside air or ground, and then they move the heat inside. In warmer months, they reverse the process to cool. There are popular types such as air-source and ground-source (geothermal) heat pumps that are really effective in reducing the energy costs and the environmental damage.

Figure 1.18: Heat Pumps

Source: (sprsunheatpump, 2024)

2. Furnaces

Heating homes and buildings is widely accomplished with furnaces. These heaters burn fuel (such as natural gas, propane, or oil) to heat air, which is then spread through the space over ducts. Newer furnaces are designed to use less energy, with features like variable speed fans and smart controls that use less, as well as variable speed blowers.

Figure 1.19: Furnaces

Source: (Richert, 2022)

3. Boilers

Boilers heat water, and that water is then pumped through pipes to radiators or a system of underfloor heating. Older buildings use them and they are powered by natural gas, oil or electricity. Boilers are also known to provide consistent and comfortable heat and are reliable in many applications.

Figure 1.20: Boilers

Source: (Flickr, 2022)

4. In-Floor Radiant Heating

The radiant heating systems heat floors, walls or ceilings directly. There electric heating element based or water filled tubes embedded in surfaces. Radiant heating, because they don't depend on air circulation, offer even warmth and can also reduce allergens in the air, making it a popular choice for the home.

Figure 1.21: In-Floor Radiant Heating

Source: (Continal, 2020)

1.3.4 Cooling Technologies: Air Conditioning, Chillers, and Cooling Towers

Keeping the indoor comfort would depend on cooling technologies, all the more so in hot climates or summer months. These technologies are responsible for the removal of heat from indoor spaces, which makes them cool and controlled. Cooling systems are employed in residential homes as well as in very large commercial and industrial structures. Three common cooling technologies are air conditioning, chillers, and cooling towers, used to attain different types of suitable cooling and reduce the use of energy.

1. Air Conditioning (AC)

There are two more common and known types of cooling technology used, which are the air conditioning (AC) systems. The removal works by removing the heat from the indoor air and transferring it outside. Refrigerants, chemicals that absorb heat, are used with a slew of components that vary depending on the system (evaporators, compressors, and condensers) to cycle the refrigerant between liquid and gas phases to exchange the heat.

Types of Air Conditioning Systems:

- Window AC Units are small, single-room units that work on a self-contained basis and are placed into the window.
- A split system is made up of an indoor unit (evaporator) and an outdoor unit (condenser). More efficient systems are commonly used in residential as well as commercial areas.
- Central Air Conditioning: The air is distributed using a network of ducts within buildings. The device is designed for room use and larger homes or commercial buildings.

Advantages: Air conditioning systems give specific temperature control, which enhances indoor convenience in significant warmth. Furthermore, they assist in maintaining clean air and keeping allergens at a minimum, and also reduce allergens and improve your health conditions within your home.

Disadvantages: AC systems are not energy-intensive; hence, they have lower operating costs. Another thing is that they need regular maintenance, like cleaning filters and refilling refrigerant levels.

2. Chillers

These are large industrial cooling systems that lower the temperature of water to then circulate it through a building, taking up the heat. On the other hand, Chillers are mainly used in commercial, industrial, and large-scale residential buildings where central cooling is required. Chillers come in various types of power: electricity, natural gas or steam.

Types of Chillers:

- **Air-Cooled Chillers:** These chillers use air to make this process occur. Typically, air-cooled chillers are used in smaller systems or buildings that have very little available space.

- **Water-Cooled Chillers:** These Chillers use water to transfer the heat from the refrigerant. Usually, cooled chillers are more energy efficient and are good for big buildings or industrial usage.
- **Absorption Chillers:** These operate on a heat source (such as steam or hot water) instead of electricity. Because they are commonly used in systems where waste heat is available or in environmental conscience applications.

Advantages: Chillers have the advantage of very efficient cooling for large buildings and industrial processes. High cooling loads can be handled by them and can be run in combination with other cooling technologies, including air conditioning or cooling towers.

Disadvantages: Chillers are difficult and expensive to install and operate. Apart from that, they use water and need a cooling tower (in water-cooled) system to function well.

3. Cooling Towers

Water used to cool air conditioning and industrial cooling systems is cooled in cooling towers. The water is heated in the reaction that takes place in the tower, is circulated through the tower, is exposed to air, and cools down by evaporation. Then the cooled water recirculates through the system to keep at the desired temperature.

Types of Cooling Towers:

- The process of evaporation is used by these towers to cool the water. For large systems requiring the dissipation of high amounts of heat, more common and efficient wet towers are used.
- Water and air do not come into direct contact in dry cooling towers. But instead, water is separated from the air through a heat exchange process to cool the water. Dry towers are preferred when there is a water shortage or when water conservation is an important consideration.

- Hybrid Cooling Towers are those that employ both wet and dry cooling processes at once, thus combining both advantages based upon the environment.

Advantages: Large-scale applications: Cooling towers are very efficient and cost-effective. Their ability to handle a large amount of cooling load makes them suitable for use in industrial, commercial, and HVAC applications. In addition, they provide water conservation benefits as well when coupled with water treatment systems.

Disadvantages: It is essential to have a large space and infrastructure in cooling towers. There is also a need for maintenance because if the water is not contaminated with minerals or bacteria, it can be a health risk. (Stanford III, 2016).

1.3.5 Basics of Ventilation and Air Distribution

HVAC systems involve ventilation and air distribution components that are critical in ensuring indoor air quality, comfort, as well as safety by developing an appropriate balance between the amount of incoming delivered fresh air and the amount of recirculating air in the building. Ventilation is proper and will deliver fresh air for indoor spaces while at the same time removing contaminants, excess moisture, and pollutants. The term air distribution is the efficient delivery of conditioned air (heated, cooled, or ventilated) in a building to achieve uniform temperature and air quality.

Practically, an HVAC designer should understand the basics of ventilation and air distribution to better understand how to design and maintain appropriate energy efficiency of the system while ensuring occupant comfort. Proper airflow and ventilation provide a means to better control the inside buildup of detrimental particles, including volatile organic compounds (VOCs), carbon dioxide, and particulate matter with an adverse impact on health, amongst other things. The quality of the distribution of air is also effective in ensuring that rooms receive enough heating or cooling, for example, so that they can be maintained in a comfortable environment.

Ventilation

Several ventilation systems bring exterior air into buildings while spreading it throughout the rooms and building spaces. The overall goal of ventilation systems in buildings should deliver clean breathing air through two main functions, which include dilution of indoor pollutants and pollution elimination. (Mărgulescu et al., 2014).

The essential aspects of building ventilation consist of three major components.

- **Ventilation Rate** — Ventilation rate stands as the essential factor regarding the external air supply volume entering building spaces, alongside outside oxygen purity evaluation.
- **Airflow Direction** — The complete airflow direction in buildings must follow a route that takes outdoor air from clean spaces to more contaminated areas.
- **Air Distribution or Airflow Pattern** — The airflow pattern should allow external air to reach every area of the space effectively, and the internal pollutants to exit the space similarly effectively.

Buildings can be ventilated through three methods that include natural ventilation and mechanical ventilation together with hybrid (mixed-mode) ventilation.

Air Distribution

Two-thirds of the residential structures in the United States, covering low-rise townhouses along with condos, utilize duct systems through central heat pumps and fans for heating and cooling energy distribution. The purpose of a forced-air system is to distribute air throughout the house, but ventilation refers to managing the entrance and exit points of indoor air. Most building infiltration stems from the combination of your ventilation ducts and air distribution system. An air handler, named a furnace fan, distributes heating and cooling air, but it carries additional air beyond its intended quantity. Every duct installation releases air due to its natural tendency to leak. A large portion of homes store their central equipment along with leaky supply and return ducts in uninsulated attic areas that allow full air transfers between indoor and outdoor spaces (SmarterHouse, 2024).

Types of Ventilation Systems

- **Natural Ventilation**

Opening such as windows, vents or louvers that allow air to flow through a building without mechanical aid are used for natural ventilation. This method is especially useful in mild climate and low occupancy spaces. For example, residential homes employ strategically placed windows and roof vents to achieve cross ventilation so that air flows without additional use of energy.

- **Mechanical Ventilation**

With large buildings, natural ventilation is not sufficient, thus mechanical ventilation is the use of fans, ducts and air handling units for controlling airflow. An example of this is a commercial office that may have a centralized HVAC system with energy recovery ventilators (ERVs) to introduce fresh air and minimize energy loss.

- **Hybrid Ventilation (Mixed-Mode)**

Hybrid ventilation is a combination of natural and mechanical systems for flexibility. Only when natural ventilation is not enough does mechanical systems activate. Operable windows in schools are an excellent example; operable windows provide ventilation when the weather is moderate, the mechanical systems provide airflow when the weather is extreme.

<u>Real-World Applications:</u>

- **Residential Buildings**

Energy efficient windows and exhaust fans in kitchens and bathrooms are often included in efficient ventilation in homes. For example, California's multi family housing project installed demand controlled ventilation (DCV) systems that adjust the airflow depending on the occupancy. The energy use was reduced by 20%.

- **Commercial Buildings**

Zoned ventilation systems with the advanced controls are used in large office spaces in managing the airflow on the basis of occupancy and temperature. For example, an IoT enabled ventilation in a high-rise office adjusted real time parameters depending on the air quality sensors. The strategy also added on to improve the indoor air quality (IAQ) and saved 15 percent on energy costs.

- **Industrial Buildings**

When it comes to dust, fumes, and excessive heat, factories and warehouses need strong ventilation. In Detroit, an automotive plant used a displacement ventilation system that delivered cool air at the floor. By removing pollutants and reducing cooling costs by 30%, this design was effectively accomplished.

1.4 Energy Consumption in Buildings and HVAC Systems

As both these are critical factors in operational costs and environmental impact, energy consumption in buildings is a critical factor. A large amount of global energy is used by buildings, especially commercial and residential ones. The use of HVAC systems accounts for a significant portion of this energy usage. Energy consumption by HVAC systems accounts for around 40-50% of a building's total energy consumption, which fits well as a part of efforts to reduce energy and carbon footprints. It is necessary to understand the consumption of energy in buildings and how HVAC systems are part of the process.

1. Energy Consumption in Buildings

Some things influence the amount of energy a building consumes, including its size, the use for which it is intended, how well the building is insulated, and how the building's systems operate. Electricity, natural gas, or renewable energy sources are typical for powering buildings, and energy consumption is shared among divisions, for example:

- **Space Heating and Cooling:** Building energy consumption by far surpasses space heating and cooling for the majority of buildings. In the cold climate, space heating is the main consumer of energy, while in the warm climate, the air conditioning systems are the main users of energy.
- **Lighting:** Another significant energy user is lighting, which is one of the leading energy users, mainly in large commercial and industrial buildings with lights on for long hours.
- **Appliances and Electronics:** Electrical appliances and electronic devices: Electrical appliances and electronic devices including computers, refrigerators and machinery also contribute to the overall energy consumption of a building.
- **Water Heating:** Other significant elements of a building's energy profile include Water Heating, where it's used for heating water (for showers, dishwashing and other domestic purposes).
- **Energy Efficient Building:** Contemporary building designs have adopted various energy-efficient techniques to reduce energy consumption, including high-performance windows, better insulation, energy-efficient lighting, and energy recovery systems. All of this helps in lowering the total demand for heating and cooling and, consequently, the load on HVAC systems.

Real World Example:

An office building, which is 30 years old, located in New York, was retrofitted to increase its energy efficiency of HVAC (heating ventilation and air conditioning) systems. This was to install highly efficient heat pumps and a Building Management System (BMS) for real-time monitoring. After retrofit, the HVAC energy consumption of the building was decreased by 35% and annual energy cost savings of $500,000 in energy were achieved with improved indoor air quality.

A solar-assisted cooling system was integrated into the HVAC setup of a luxury hotel in Dubai. During the day, solar panels supplied the energy for cooling, considerably decreasing demand on conventional power sources. By having this innovation, the hotel was able to reduce overall energy consumption by 25% and was also on par with local green building regulations.

2. Energy Consumption in HVAC Systems

HVAC systems are the largest energy-consuming systems in the building; they are designed to keep the environment in the building comfortable, but at the same time, HVAC systems use the most energy in the building. Because they continuously run, regulating temperature, humidity, and air quality in commercial buildings, their role includes. According to the type of HVAC systems, the energy consumption can be grouped as below.

- **Heating:** Boilers, heat pumps, or furnaces that provide heating system: these heat systems utilize a lot of energy to create indoor comfort in the colder seasons. Energy consumption depends greatly on the type of heating system, its age, the efficiency rating, and the way it has been maintained. Large amounts of energy are lost with older or poorly maintained systems.
- **Cooling:** Air conditioning systems, chillers, and cooling towers are used to cool the air inside in warmer periods of the year. In regions with hot and long summers, air conditioning units eat a lot of energy, as they do with heating systems. Cooling systems are efficient when a refrigerant is used, the unit's energy efficiency rating (EER or SEER), and the system's maintenance are considered.

3. Renewable Energy Integration

In pursuit of sustainability, a large number of building systems are being made to incorporate renewable energy sources (RES), such as solar power, to balance the amount of energy consumed by HVAC systems. Especially during the high-demand period of summer, solar panels can be used to power HVAC systems. In addition, solar water heating systems help cut down on the use of traditional electric water heaters.

4. The Role of Energy Management Systems

Real-time monitoring and control of building energy use is done by energy management systems (EMS) and building energy management systems (BEMS). These systems collect data on energy consumption and then optimize the HVAC operations with real-time conditions, weather forecasts, and occupancy patterns. The integration of energy-efficient best practices with smart building technology can significantly reduce energy wasted, financial cost, and environmental degradation, which are caused by both EMS and BEMS.

Real World Examples

Advanced airflow management and variable refrigerant flow (VRF) systems deployed in a data center in California reduced and optimized cooling. This resulted in a reduction of Power Usage Effectiveness (PUE) of the facility from 1.9 to 1.4, proving a solid energy efficiency while maintaining reliable operations.

1.5 Chapter Synthesis

In this chapter, the main insight from different heating, cooling, and ventilation technologies is synthesized, and their critical role in building energy management is demonstrated. Heating systems from heat pumps, furnaces, boilers, and radiant heating have different efficiency levels, cost-effectiveness, and suitability depending on house quality as well as climate. Cooling technologies, such as air conditioners, chillers, and cooling towers, are important for comfort in warmer climates, but these technologies also account for a massive proportion of energy consumption. Ventilation and air distribution systems involve ventilation and air distribution and are in charge of preserving optimal environmental conditions with acceptable air quality. Nowadays, the integration of RES such as solar power in HVAC systems is increasingly critical to decouple the use of HVAC systems from conventional energy and carbon footprints. The implementation of energy management systems (EMS) also has the effect of supplying one way of real-time control and optimization of building energy usage to secure a reduction of cost and improve the environmental impact. In the end, efficient heating, cooling, and ventilation technologies, along with renewable energy and energy management systems, can have a substantial influence on the performance and sustainability of the building.

Multiple Choice Questions (MCQs)

1. **What is the primary goal of energy efficiency in HVAC systems?**

    a. Increase energy consumption
    b. Maintain comfort levels while reducing energy use
    c. Enhance system complexity
    d. Reduction of initial system costs

2. **Which professionals are primarily responsible for implementing energy conservation measures in HVAC systems?**

    a. Architects
    b. HVAC technicians
    c. Engineers and facility managers
    d. Building occupants

3. **Which of the following is a characteristic of centralized HVAC systems?**

    a. No ductwork required
    b. Single, central unit serving multiple zones
    c. Individual units for each room
    d. Limited control over individual room temperatures

4. **Which component is responsible for distributing conditioned air in an HVAC system?**

    a. Variable Air Volume (VAV)
    b. Chiller
    c. Ventilation Ductwork
    d. Air Handling Unit (AHU)

5. **What is the primary function of boilers in HVAC systems?**

    a. Filter indoor air
    b. Regulate humidity levels
    c. Cool indoor air
    d. Heat water or air for heating purposes

6. **Which technology is commonly used for cooling in HVAC systems?**

    a. Boilers
    b. Radiant heating
    c. Heat pumps
    d. Air conditioners

7. **What does the term 'ventilation' refer to in HVAC systems?**

    a. Distribution of conditioned air
    b. Exchange of indoor and outdoor air to maintain air quality
    c. Cooling indoor spaces
    d. Heating indoor spaces

8. **Which of the following is a decentralized HVAC system component?**

    a. Chiller
    b. Cooling tower
    c. Air Handling Unit (AHU)
    d. Variable Refrigerant Flow (VRF) system

9. **How does energy consumption in buildings relate to HVAC systems?**

    a. HVAC systems consume energy only during extreme weather conditions
    b. HVAC systems are the primary contributors to energy consumption in buildings
    c. Energy consumption is unrelated to HVAC system efficiency
    d. HVAC systems have minimal impact on building energy consumption

10. **Which of the following is a key challenge in HVAC energy optimization?**

   a. Eliminating the use of renewable energy sources
   b. Reducing the size of HVAC equipment
   c. Increasing system complexity
   d. Balancing energy savings with occupant comfort

Answers

1	2	3	4	5	6	7	8	9	10
b	c	b	c	d	d	b	d	b	d

CHAPTER 02

ENERGY EFFICIENCY STRATEGIES IN HVAC SYSTEMS

2.1 Chapter Overview

This chapter elaborates upon a view of the evolving scene of building management across smart Heating, Ventilation, and Air Conditioning (HVAC) systems, Building Automation Systems (BAS), and the shaping function of IoT and AI technologies. Here, the paper covers smart HVAC systems by detailing how they integrate IoT components to optimize temperature control, increase energy efficiency, and enhance indoor air quality at the initial stage. The chapter continues with a discussion of the components and benefits of BAS, i.e., centralized control of multiple systems in the building for efficiency and comfort. It goes on to discuss how IoT has contributed to HVAC systems, where smart thermostats, occupancy sensors, as well as air quality monitoring were applied, and then it discusses the issue of security, data overload, and the costs. The integration of AI and machine learning (ML) in HVAC systems is then described, wherein its role is defined in terms of prediction of system behavior, performance optimization, and predictive maintenance. The main point of the chapter revolves around these technologies, which are changing the face of building management towards smarter, more sustainable, and more cost-efficient building management.

2.2 Load Estimation and System Sizing

Estimation of load and sizing of the system are vital considerations in the design of an energy-efficient HVAC system. The adequate heating, cooling, ventilation and minimizing of energy consumed by a system is ensured by appropriate load estimate. The proper sizing of the HVAC system is critical to avoid it being too small and thus underpowered and create inefficiency and unnecessary costs in the operation or being too large and thus overwhelming and also inefficient and costly to operate.

Factors Influencing Load Estimation

The thermal and ventilation requirements of a space can be assessed during the load estimation process. The following factors play a very important role in deciding the heating and cooling loads of a building.

- **Building Insulation and Envelope:** The quality of insulation, windows, and construction materials has a direct bearing on the amount of heat that a building gains or loses. Such well-insulated buildings decrease heating and cooling loads, resulting in more energy savings.

- **Internal Heat Load, Occupancy, and Activity Levels:** The number of occupants and the schedule of activities performed within a space greatly affect the internal heat load and, to some degree, the ventilation requirements. More cooling is required when it is in high-density areas or in spaces where equipment-generating heat is present.

- **External Weather Conditions:** The heating and cooling demands for a building are dependent on external weather conditions, including air temperature, humidity, solar radiation, and wind. For accurate calculation, that information is the local climate.

- **Internal Heat Gains:** Account should also be taken of Internal Heat Gains, which are the amounts of heat created by lighting, office equipment, machinery etc. within the building. This internal load is usually larger in commercial buildings with large equipment use.

- **Ventilation Requirements:** The proper air quality requires good ventilation. Taking into account that there are air changes per hour (ACH) is a case factor determining the ventilation load on the HVAC system and total cooling and heating needs(tutorials point, 2024).

Load Calculation Methods

The various ways in which heating and cooling loads can be calculated include simple empirical and more complicated methods based on simulation. Commonly used methods include:

- **Manual J Calculation (Residential):** This is a simplified procedure to estimate heating and cooling loads in residential buildings. It assumes building size, insulation, windows, occupancy, etc. to come up with system requirements.
- **American Society of Heating, Refrigerating, and Air-Conditioning Engineers (ASHRAE) Methods (Commercial and Industrial):** In the case of commercial and industrial buildings, ASHRAE provides specific guidelines for the calculation of load. Building orientation, material, occupancy, and climate data are used that have been integrated into ASHRAE methodology to estimate heating and cooling loads.
- **Energy Simulation Software:** This includes advanced tools such as Energy Plus and TRNSYS for energy performance simulation of the building. The fact that complex variables such as air infiltration, energy usage patterns, and performance under a variety of conditions are factored into these tools leads to a more detailed analysis(Burdick, 2012).

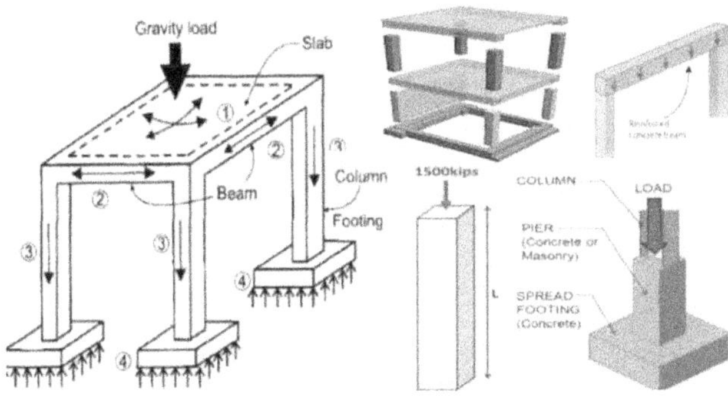

Figure 2.1: Load Estimation

Source: (KG, 2023)

System Sizing

The next step is to properly size the HVAC system after the load assessment is finished. While an enormous system will cycle on and off frequently, decreasing efficiency and increasing wear and tear, an undersized system will find it difficult to maintain the ideal interior conditions. Key considerations in system sizing include:

- **Capacity Match:** The HVAC system should be sized to match the peak load of the building, which occurs under the most demanding conditions (e.g., the highest outdoor temperature). However, the system should also have the flexibility to operate efficiently during lower-demand periods.

- **Efficiency Ratings:** The sizing process should take into account the energy efficiency ratings of the system, such as Annual Fuel Utilization Efficiency (AFUE) for heating and the Seasonal Energy Efficiency Ratio (SEER) for cooling. A higher efficiency rating reduces energy consumption and long-term costs.

- **Future Load Considerations:** It is important to account for potential changes in building occupancy, usage, or design. Future modifications to the space (e.g., added rooms or equipment) should be considered to prevent the need for a system upgrade.

- **Load Diversity:** In commercial and industrial applications, not all areas of a building will reach peak load conditions simultaneously. Therefore, a diverse load factor can be applied to reduce oversizing while ensuring comfort in all areas.

Optimizing Load Estimation and System Sizing

Assessment of load should come naturally with building design strategies and technological innovations, to maximize energy efficiency at the best possible location.(EDS, 2024). Some approaches include:

- **Real-time Conditions Monitoring:** Such sensors can monitor the real time conditions like temperature, humidity and occupancy (e.g., as the number of people in a room) and can automatically adjust the HVAC system based on them. It neglects the needs but reduces energy consumption by changing the system operation according to the needs.
- **Demand response programs:** Joining one of these demand response programs will help to optimize HVAC unit load at peak demand times, helping both the building owner and the broader grid.
- **HVAC system and Variable Air Volume (VAV) Systems:** By zoning the HVAC system and VAV system, we could control the airflow more efficiently and lower the load on the system by locating the area in the building itself rather than running a constant through the building.

2.3   HVAC System Design and Optimization

The integration of HVAC systems remains essential because they provide comfort standards in buildings through ventilation and heating needs, and sustainable air management. The design process, along with optimization systems, results in an optimal indoor climate and simultaneously minimizes operational costs, reduces energy usage, and supports sustainable practices. The fundamentals and strategic methods used to create optimal HVAC systems for these goals are presented in this part.

Principles of HVAC System Design

1. **Load Calculation:** Organizations need precise load calculations because they determine both the replacement size of HVAC equipment and its operational capacity. Heating and cooling demand estimation for buildings depends on climate conditions in combination with occupancy patterns along with insulation standards of the envelope and directional orientation and the number of internal heating sources.

2. **System Sizing and Selection:** An inappropriate size of HVAC systems will result in either poor energy conservation or inadequate performance. Alignment between system size and its operational boundaries enables maximum efficiency, thus minimizing power usage and maintenance expenses.

3. **Component Selection:** Quality energy efficiency depends on selecting components known to minimize power use. The installation of high-efficiency elements such as boilers plus chillers, pumps, and fans provides substantial operational expense reduction for HVAC systems.

Optimization Strategies for HVAC Systems

1. **Variable Speed Drives (VSDs):** The implementation of VSDs enables HVAC system motors to adapt their speed according to demand, which results in decreased energy consumption when system loads are low. The installation of Variable Speed Drives (VSDs) in HVAC components should allow controllers to change airflow and water flow rates to optimize energy consumption under changing occupancy scenarios and environmental conditions.

2. **Energy Recovery Ventilation (ERV):** EROs function as energy recovery devices that extract heat from exhaust flows to warm or cool ventilating air, thus preparing it for HVAC system entry. Such

a setup alleviates heat stress on heating as well as cooling systems to deliver better efficiency of the entire system.

3. **Demand-Controlled Ventilation (DCV):** DCV systems vary ventilation rates according to occupancy, such that energy is wasted by the HVAC system only when the spaces are unoccupied or under-occupied. Reduce energy use in cases of low occupancy by actually implementing CO2 sensors or occupancy sensors to optimize ventilation rates.

4. **Building Automation Systems (BAS):** BAS is important as they are good for central monitoring and control of HVAC equipment, lighting, and other systems. Automating the control of these systems based on real-time data allows these systems to operate their HVAC systems optimally and use energy optimally. Use it to continuously monitor the environmental variables such as temperature, humidity, occupancy, etc., and other possible variables, and make real-time changes to the HVAC settings to minimize the energy used.

5. **Zoning:** Zoning divides a building into separate areas that can be conditioned independently so that energy is used where and when it is needed. Zoning cuts down on the need to have the entire HVAC system run at full utility. In order to control the temperature and airflow in the different boundaries of a building (conference rooms, offices, or hallways), these boundaries can be separated by using Variable Air Volume (VAV) systems and smart thermostats with separate setpoints for temperature and airflow.

2.3.1 Best Practices for Designing Energy-Efficient HVAC Systems

To maximize the building's energy efficiency and sustainability gains, a thorough analysis is needed for designing the energy-efficient HVAC system which includes reviving accurate load calculations, correct system sizing and integration of the latest technologies. These best functions not only guarantee that the building is comfortable, but they reduce energy consumption, lower operational costs and enhance sustainability. The following are additional detailed strategies and recommendations for designing energy-efficient HVAC systems.

1. Accurate Load Calculations and Customization

- **Importance:** Proper load estimation means that the HVAC system will be sized correctly and the HVAC system will run efficiently. Calculation of load results in energy waste and improper system performance if the load is inaccurate.
- **Best practice:**
  - **Dynamic Load Calculations:** Using software such like Energy Plus or TRACE 700 to consider real-time changes in how the building is used, weather, and occupancy patterns. These tools help model the energy use when the building is operating, including the influence of demand fluctuations.
  - **Internal Gains:** Heat from equipment, lighting, and occupants as internal gains should be considered in the load calculation. This will prevent over-sizing or under-sizing of the HVAC system.
  - **Daylight Savings and Occupancy Changes:** Incorporates variations in occupancy, and solar heat gain across different times of day schedules as well as different seasons, to allow flexible changes such as reducing unnecessary heating or cooling.

2. Optimizing System Sizing

- **Why it's important:** If the size of HVAC equipment is incorrect, the bad substances include high energy use, poor humidity control, and undue wear and tear on the system.

- **Best practice:**

  - **Refuse Oversize:** While selecting equipment, always select a capacity higher than the peak load to avoid oversizing. Troops turn on and off more frequently, thus causing energy inefficiency.

  - **Variable Capacity Systems:** Select a modulating system or variable refrigerant flow (VRF) system that can change the output according to actual demand, and, therefore, the energy consumption would be more critical.

  - **Include future load adjustments:** Include building expansions, potential improvements in technology, or more future insulation improvements that may allow the system to adapt to the changes without needing additional recalibration(Tosin Michael Olatunde et al., 2024).

2.3.2   Role of Building Orientation, Insulation, and Envelope

The building's orientation, insulation, and envelope design all have a big impact on how energy-efficient HVAC systems are. These architectural elements can either enhance or hinder the effectiveness of HVAC systems in regulating temperature and maintaining comfort levels while reducing energy consumption. A thoughtful integration of these factors can lead to more sustainable and cost-effective heating and cooling solutions. Below is a breakdown of how each of these aspects contributes to HVAC energy efficiency:

1. Building orientation refers to how a structure is oriented concerning the paths of the sun during the various seasons and the patterns of the predominant winds. The design and placement of living and sleeping spaces, whether to benefit from the sun and wind or to be shielded from their impacts, is another aspect of passive design. For instance, the building is oriented North-South if its length is East-West. In terms of comfort in the natural

elements, proper orientation refers to establishing or adjusting the directions of the building's plan so that the occupants can enjoy nearly all that is good and stay away from everything bad. (Technical Editor, 2024).

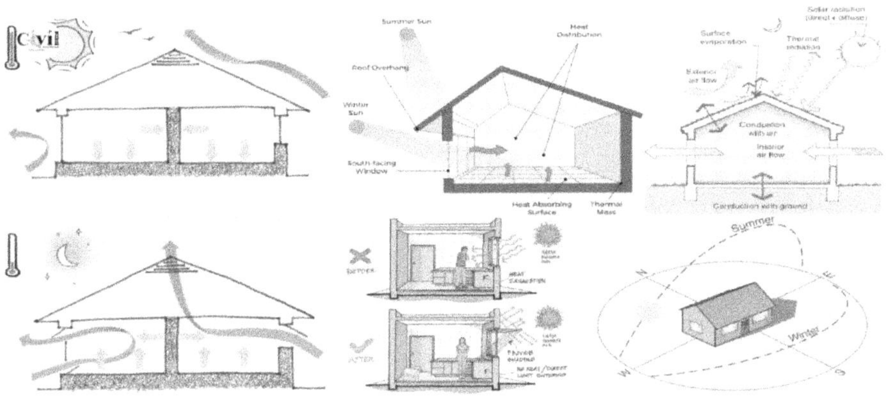

Figure 2.2: Building Orientation

Source: (Discoveries, 2024)

Principles of good orientation

- When paired with other energy-saving measures, good orientation can minimizer even completely do away with the need for auxiliary heating and cooling, which lowers energy costs, lowers greenhouse gas emissions, and improves comfort.
- It considers changes in the sun's course from summer to winter, as well as the type and direction of winds, including cooling breezes.
- The use of auxiliary heating and cooling systems can be significantly reduced or eliminated with proper orientation. This leads to increased comfort, lower energy bills, and a decrease in greenhouse gas emissions.
- Aim to block direct sunlight by shading every façade all year round with trees and nearby buildings in hot, humid climates or

hot, dry regions that don't need winter heating. Capture and direct cooling breezes.

- In winter, the sun's rays reach north-facing surfaces at a higher rate than in summer. (Discoveries, 2024).

2. **Building Insulation:** The term "thermal insulation" describes the materials used to keep a building at a constant, comfortable temperature. These materials are called thermal insulators. Thermal insulation is the practice of keeping heat from escaping a building. Reduced heating and cooling demands mean lower utility bills and a higher quality of life thanks to well-insulated buildings.

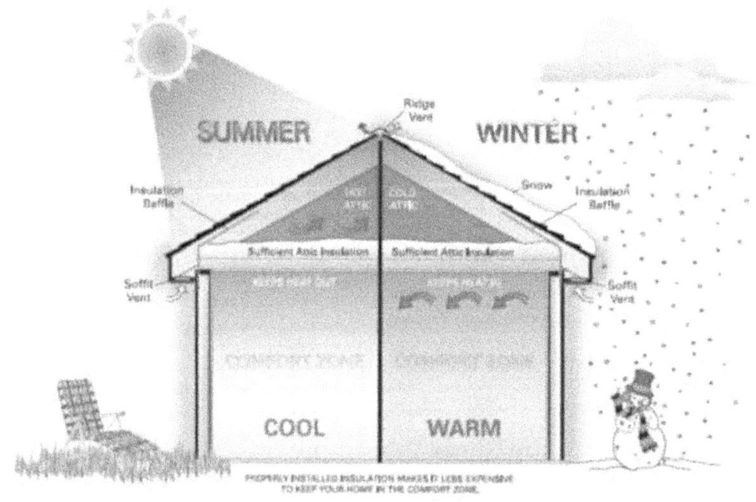

Figure 2.3: Building Insulation

Source: (dailycivil, 2024)

Reducing energy use for heating and cooling buildings has environmental benefits. A building's thermal insulation should work to lessen the transfer of heat from the outside to the inside. In addition to reducing noise, preventing fires, and increasing a building's lifespan, thermal insulation materials save energy. Consequently, thermal insulation has grown in importance in contemporary building practices.

Types of Thermal Insulation

Building thermal insulation comes in many forms, including the following:

- **Bat Insulation:** Insulation materials for walls and ceilings can be found in blanket form or in the form of paper rolls. Their thickness can vary from 12 to 80 mm, and they are both flexible and strong. A variety of materials, including cotton, wood fibers, and animal hair, go into making these blankets.

Figure 2.4: Bat Insulation

Source: (dailycivil, 2024)

- **Slab or Block Insulation:** Insulators in the form of blocks or slabs are small, immovable objects with dimensions of 60 by 120 cm and a thickness of 2.5 cm. Some of the materials used to manufacture the blocks include sawdust, cellular rubber, mineral wool, corkboard, and cellular glass.

Figure 2.5: Slab or Block Insulation

Source: (dailycivil, 2024)

The ingredients are united to form blocks using cement. It is possible to line the walls and roofs with these small pieces. These are fastened to the roofs and walls to keep the heat in and the temperature where it needs to be.

- **Insulating Boards:** Insulating boards can be made from a variety of materials, including wood pulp and cane pulp. To make these boards, the pulp is heated to the right temperature and then compressed with a certain amount of force. They are commonly supplied for use as interior wall lining and partition walls, and they come in a range of sizes on the market.

Figure 2.6: Insulating Boards

Source: (dailycivil, 2024)

2.3.3 Importance of Thermal Comfort and IAQ Considerations

Features of a healthy, productive, and comfortable indoor environment include thermal comfort and indoor air quality (IAQ). The elements discussed here are no less important for the building occupants' well-being, but also have a direct effect on the energy efficiency of Heating, Ventilation, and Air Conditioning (HVAC) systems. Aslanoglu and coworkers (1986) emphasize that both thermal comfort and IAQ considerations are properly addressed to ensure that HVAC systems operate optimally, minimally consuming energy while maintaining a comfortable and wholesome indoor climate. In the following, we discuss the importance of these factors and their connection with HVAC performance and energy efficiency.

Thermal Comfort

It is necessarily the condition of a person who feels content with the surrounding temperature, humidity, and airflow. To achieve optimal thermal comfort, the HVAC units need to be avoided from overworking. Factors influencing thermal comfort include:

1. **Temperature Control:** Thermal comfort very much depends on a well-regulated temperature. HVAC systems that maintain the steady indoor temperature within a comfortable range (typically, between 68°F to 72°F or 20°C to 22°C for most people) prevent any drastic changes in indoor climate, therefore minimizing energy waste due to constant adjusting. Zoned systems (allowing different temperature settings in different parts of a building) also enhance energy efficiency by allowing different temperature control to different parts of a building and, with it, keeping the temperature down or up as necessary in unoccupied areas, resulting in less heating or cooling than would otherwise be required.

2. **Humidity Control:** Thermal comfort depends hugely on humidity. Humidity: High humidity can make a room feel warmer than it is, while low humidity can bring about the air feeling as though it were to have been cooler and might cause discomfort. Humidification and dehumidification in an HVAC system often serve either find the optimal indoor humidity level (typically between 30% and 60%) to increase comfort and reduce system load by preventing overcooling or overheating.

3. **Air Movement:** It is known that the movement of air in a space can impact perceived temperature. To achieve that, the airflow has to be proper to distribute heat equally and not create the impression that the environment is stagnant and stuffy. An effective air

circulation venting system that provides the required temperature, but with less reliance on an HVAC system for cooling, reduces energy consumption.

4. **Personal Preferences:** Different preferences for thermal comfort exist for the occupants, which are determined by clothing, metabolic rate, and personal sensitivity to temperature. Such flexible HVAC systems also ensure optimized energy use and occupant needs in different zones of a building.

Indoor Air Quality (IAQ)

Indoor air quality is defined as the condition of the indoor air and its effects on the health and comfort of its occupants, and, for many, the productivity, as well as the productivity and these effects often are related to building design, construction, layout, furnishings, the percentage and type of exterior walls, building presence of people and equipment within the building, processes and other variables. Poor IAQ can be associated with respiratory problems, allergies, and other symptoms that lead to reduced productivity and poor health. The consideration of IAQ is critical for energy-efficient HVAC systems to avoid excessive ventilation that is wasteful of energy.

1. **Ventilation:** For good IAQ, the air must be properly ventilated. HVAC systems with good airflow abilities supply fresh air to the building and remove stale air along with pollutants. For example, "energy recovery ventilators" (ERVs) and "heat recovery ventilators" (HRVs) are mechanical ventilation systems that transfer heat through the air streams that are entering and exiting the building to minimize the energy required for heating or cooling while still providing adequate ventilation. They offer a continuous flow of fresh air, which does not require any compromise in energy efficiency.

2. **Pollutant Control:** Pollutants that affect IAQ are volatile organic compounds (VOCs), particulate matter, mold, and carbon dioxide

(CO_2) levels. Advanced filtration and air purification technologies can be incorporated into HVAC systems, reducing to an equal extent the concentration of the above-mentioned pollutants. The combination of HEPA filters and activated carbon filters will help capture fine particles and gases and help improve IAQ, as well as ensure your HVAC system is not just about maintaining comfort, but rather protecting the health of those inside.

3. **CO_2 Monitoring:** Indoor high concentration of CO_2 is indicative of poor ventilation and could cause discomfort, fatigue and impaired cognitive performance. HVAC systems that combine CO_2 sensors with automatic ventilation adjustments according to occupant levels ensure the optimal balance between HVAC system energy utilization and IAQ. Demand-controlled ventilation (DCV) systems, on the other hand, adjust the quantity of fresh air fed into the building according to real-time CO_2 levels so that their rates of ventilation do not exceed the actual requirements of the occupants.

4. **Moisture Control:** Indoor air with excessive moisture can be a food for the further growth of mold, mildew, allergens, etc., which decreases the IAQ. HVAC systems, including dehumidification, can prevent excessive moisture buildup, helping to decrease the risk of mold and also improve the quality of air. The integration of dehumidifiers into the HVAC system and working together with cooling systems helps in controlling the humidity, improving thermal comfort, IAQ, and system efficiency in humid climates(EPA, 2024).

2.4   Optimizing Heating and Cooling Efficiency

Reduction in energy consumption, reduction in the cost of operation, and also of environmental impacts due to building operation can be achieved if it optimizes the heating and cooling efficiency in the HVAC system is optimized. Implementation of several other strategies and best practices will help improve the efficiency of HVAC systems without sacrificing occupant comfort. The following are very important approaches to optimizing heating and cooling efficiency:

1. Proper Sizing of HVAC Systems

An HVAC system that is too large or too small for the building will operate inefficiently and will consume more energy than needed. Proper sizing will make the entire system capable of meeting the building's heating and cooling load and will not over- or underwork the system.

Optimization Strategies:

- **Conduct a Load Calculation:** To commence the program, a Load Calculation should be performed, taking into consideration such factors as building size, insulation, window orientation, occupancy, and local climate.
- **Avoid Over-Sizing:** An oversized HVAC system will be cycling on and off continuously, wasting energy. A correctly sized system will run at a more efficient temperature and a level temperature inside.

2. Zoning and Smart Controls

HVAC systems in description allow these same systems in the zones of a building to be controlled independently or separately so that only the required space is conditioned or cooled, warming the remaining spaces, and reducing energy waste. One additional feature of smart controls is programmable thermostats or sensors that alter settings with their related occupancy and environmental conditions for higher system efficiency.

Optimization Strategies:

- **Zoning Systems:** Control what temperature zones are in a building (thermostats, dampers, sensors). It assures that the heating or cooling of places with diverse associations would or patterns or different levels of occupancy at any one point.

- **Programmable or Wi-Fi-enabled thermostats:** By using programmable or Wi-Fi-enabled thermostats, you can automatically have the temperature adjusted according to schedules when you are away from home or when the external temperature changes. It minimizes energy use when there is no space utilization or during off-peak times.
- **Motion Sensors:** Place motion sensors in rooms or locations to recognize occupants and enhance or decrease heating or cooling from vacant regions where excessive energy is not required.

3. Regular Maintenance and Upkeep

Routine maintenance is a description of process of ensuring that HVAC systems are running at the most efficient performance level possible to prevent performance issues that may result in increased energy consumption for aforementioned HVAC. For optimal operation of the system, regular servicing of the system is also a necessity.

Optimization Strategies:

- **Clogged or Dirty Filters:** Filters can produce sufficient dirt to decrease airflow and system efficiency. Filters should be replaced regularly or cleaned to prevent smooth operation and to increase energy efficiency.
- **Check for leaks throughout Ductwork:** If there are poor insulation or even seals too loose in the ductwork, this can be a substantial contributor to losing energy. In general, sealing and insulating ductwork helps prevent energy loss, and conditioned air reaches its intended spaces.
- **Dust and Debris on Evaporator and Condenser Coils:** If the bottle caps, events, corrugated hose, and debris get caught up on the evaporator and condenser coils and the performance of the evaporator and condenser is hindered, it increases the amount of work the system must perform.

- **Check Refrigerant Levels:** Refrigerant levels play a huge role in the effectiveness of cooling, and using the amount you need decreases the amount of energy used. The performance will be optimized if refrigerant levels are maintained at the proper level.

4. Insulation and Building Envelope Improvements

A well-insulated building retains conditioned air better, resulting in lower loading on the HVAC systems. The building enclosure components: walls, roofs, windows, and doors can really reduce requirements for excessive heat and cooling provided the building envelope is improved.

Optimization Strategies:

- **Heat Loss Reduction:** Insulation of walls, roof, and floors should be adequate to curb heat loss in winters and heat gain in summers. The more effectively the building is in retaining indoor temperatures, the higher the insulation value (R value).
- **Double-glazed or Low E (low emissivity) Windows:** This helps minimize heat transfer and better insulates the house. Proper shading can also reduce solar heat gain, such as external blinds or curtains.
- **Windows, Doors & Ductwork:** Check for gaps around windows, doors, and ductwork and seal them up to keep air out. This will avoid pushing more air through and create additional heating or cooling demand(Mannan Rana, 2024).

2.4.1 High-Efficiency Boilers and Furnace Technologies

For maximum heating performance, the lowest energy consumption, and minimum operational costs, boosting high-efficiency boilers and furnaces is crucial. The time sense of oil and gas workers has withered away, replaced by advanced technologies and design features that are designed to maximize heat output at the minimal expense of fuel. However, these systems provide some significant advantages in both the residential and commercial setups, with consequent energy saving and environmental sustainability. In the next few paragraphs, we discuss the technologies, how these boilers and furnaces work, and the benefits they bring.

Figure 2.7: High-efficiency boiler with domestic hot water tank.

Source: (efficiencymaine, 2024)

Boilers and Furnaces

The following features can help improve efficiency:

1. **Condensing:** A condensing (also known as a condensing boiler or furnace) extracts heat from the water vapor in the combustion gases, getting more "bang for your buck" out of the exhaust before it goes up the chimney. Burner efficiencies of up to 96% can be achieved in condensing systems.

2. **Modulating Burner:** These boilers and furnaces will modulate their burner output to respond to the heating needs of the home, without cycling on and off each hour the thermostat requires additional heat. If "short cycling," ie, turning on and off frequently, a simple 'on/off' boiler may be operating inefficiently, as it now heats up and cools down.

3. **Sealed Combustion:** Sealed Combustion Systems bring outside air directly into the unit and tend to be more efficient since the combustion is not being done with indoor air. Furthermore, they lower the risk of backdrafts and carbon monoxide leaks.'

4. **Proper Sizing:** Oversized boilers and furnaces are more costly to buy, and the cycle more often than efficient ones, lowering overall efficiency (efficiencymaine, 2024).

High-Efficiency Boilers

The water that's circulated throughout your home to deliver heat… it surrounds the combustion chamber and flue passages (called the heat exchanger). Second, these flue passages, by design, harvest heat in the flue gas and feed the water of the heat exchanger. The more value you can get from every dollar you spend on fuel, the more energy the heat exchanger can harvest from the flue gas.

The process of "condensing" allows a high-efficiency boiler to repurpose the heat that would otherwise escape and use it to heat the house.

Boilers that are designed for high-efficiency condensing use fuels like natural gas or propane to generate heat. The term "condensing" describes these devices because, in comparison to regular boilers, they can significantly reduce the temperature of the flue gases as they exit the appliance. The flue gases condense within the heat exchanger due to their extremely low temperature. (Company, 2024).

Types of High-Efficiency Boilers

*   **Condensation Boilers:** Efficiency of up to 98%, almost all the fuel energy can be converted into usable heat with these boilers. For

homes and buildings with advanced heating demands, they are optimal.

- **Non-Condensation Boilers:** The boilers are now more efficient than the non-condensing models, achieving high-efficiency levels (approx. 90%) with modern heating controls and well-insulated distribution systems.

High-Efficiency Furnaces

Such furnaces are highly efficient and they work under modern technology to generate heat in maximum amount and as little energy waste as possible. Secondary heat exchangers are also present with high-efficiency furnaces. Air is warmed by the primary heat exchanger while the secondary heat exchanger captures and recovers heat from the exhaust gases, usually lost to the flue. The furnace can take more heat from the same interval of fuel because of this. Also, these furnaces are condensing models; they can capture and condense water vapor left over from the combustion process. What's more, the moisture generated from the condensation also provides heat to warm the air inside the furnace, and in doing so lowers the bill. Variable-speed blowers provide variable airflow in high-efficiency furnaces to cater to varying heating demands. In case of cold, this helps warm up air more evenly and continuously, and thus reduces energy consumption and provides comfort.

A lot of high-efficiency furnaces have modulating burners, which change the intensity of the flame based on how much heat the house needs. This implies that during milder weather, the furnace doesn't always run at full capacity, conserving energy.

The sealed combustion mechanism of high-efficiency furnaces pulls air from the outside of the house to fuel the fire. This enhances safety and efficiency by preventing the furnace from reducing the interior air. (Travis Baugh, 2024).

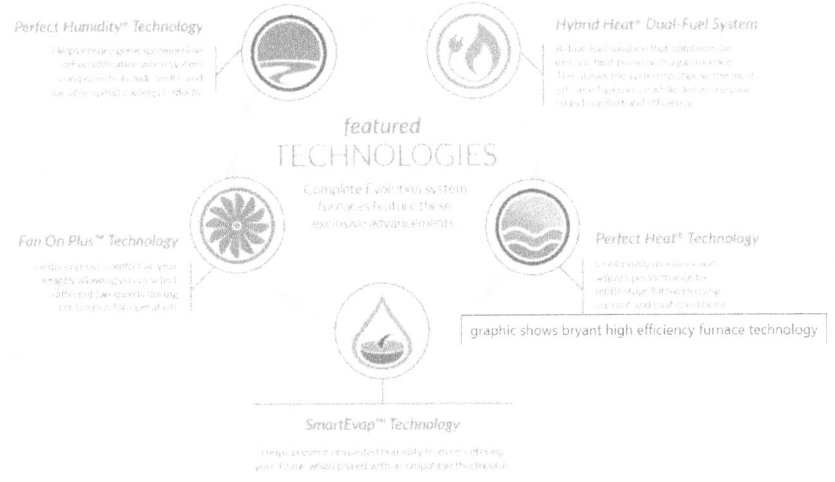

Figure 2.8: Featured Technologies

Source: (Travis Baugh, 2024)

Benefits of High-Efficiency Furnaces

For homeowners, high-efficiency furnaces have a lot to offer. They supply good heating in winter, savings on energy, and thus on utility bills. Savings can be found in the hundreds over long periods if you upgrade to a high-efficiency gas furnace.

- **Energy Efficiency:** One benefit is their energy efficiency. These furnaces have efficiency ratings usually above 90% AFUE (Annual Fuel Utilization Efficiency), which means they are more efficient at providing usable heat, using less fuel and therefore using less wasteful energy, and keeping heating bills low. The reality is that this efficiency not only helps the environment but can also be translated to noticeable energy savings as compared to older furnace models. Bryant Evolution furnaces have a high efficiency of up to 98.3 % AFUE, making them a comfortable, high-efficiency choice.

- **Lower Noise Levels:** In addition, high-efficiency furnaces have lower noise levels. These have noise-reducing features that keep your home without any constant noise noise of the conventional furnaces.
- **Environmental Impact:** High-efficiency furnaces are more eco-friendly in their use of less fuel with fewer emissions. This can help to reduce a bit of your house's carbon footprint and play a part at hand in greening your environment. Learn how Bryant is committed to sustainability.

2.4.2 Advanced Heat Pump Systems and Applications

Heat pump performance is being enhanced by some innovations:

1. Staged or Multi-Speed Compressors

- Heat pumps can function at temperatures outside close to the required heating or cooling capacity thanks to staged or multi-speed compressors.
- These systems reduce compressor wear and on/off operation, which saves energy.
- Inverter-driven systems can modulate their speed/capacity at near-infinite degrees between low and high settings, running efficiently and maintaining consistent comfort.

2. Variable-Speed or Dual-Speed Motors

- The majority of contemporary heat pump and the furnace blower types have variable-speed or dual-speed motors on their indoor fans, outdoor fans, or both, or electrically commutated motors (ECM).
- These fans' variable-speed controls minimize cool drafts, maximize power savings, and minimize noise and disturbance by maintaining the air flow at a suitable pace.

3. Desuperheater

- A desuperheater, a feature of some high-efficiency heat pumps, recovers waste heat from the cooling phase of the heat pump and uses it to heat water.
- The efficiency of a heat pump with a desuperheater is two to three times higher than that of a standard electric-resistance water heater.

4. Dual-fuel or Hybrid Systems

The dependability of a gas furnace and the efficiency of a heat pump are combined in dual-fuel or hybrid systems. This combination allows the heat pump to handle most of the heating needs in milder weather, while the furnace takes over during colder temperatures.

How It Works:

- **Energy Efficiency:** During warmer months, the heat pump efficiently heats and cools your home. When temperatures drop, the system automatically switches to the gas furnace, which is better suited for cold weather.
- **Shared Ductwork:** Both systems typically use the same ductwork, making it a straightforward installation if you're upgrading from a traditional furnace and air conditioning setup.
- **Widely Available:** Commonly found in regions like the Mid-Atlantic, dual-fuel systems are easy to implement and help homeowners reduce electricity use in cold climates while maintaining comfort year-round.

This setup is a great option for homeowners looking to maximize energy savings while ensuring reliable heating in colder weather.

5. Cold Climate Heat Pumps

- Heat pumps for cold climates are made to function in temperatures as low as 5°F.

- If you live in a region where temperatures regularly dip below freezing in winter months, consider looking for a system with an Energy Star Cold Climate label.
- Learn more about Energy Star Cold Climate heat pumps.

Applications of Advanced Heat Pump Systems

a. **Residential Heating and Cooling**

- **Heat pumps** are an effective alternative to furnaces and air conditioners because they allow homes in temperate to cold climates to stay hot in the winter and cool in the summer.
- **Voice assistants and mobile apps integration:** Integration of voice assistants such as Siri or Alexa and mobile apps, such as the app on the phone, can be integrated with this heating and cooling schedule for their convenience.

b. **Commercial and Industrial Applications**

- Businesses and office spaces: Heat pumps provide energy-efficient heating and cooling services for businesses and office buildings, hence lowering the operating costs and the impact on the environment.
- **Industrial applications:** Heat pumps can be used in industrial applications for space conditioning and process heating, which; constitutes a reliable and economic alternative.

c. **District Heating and Cooling Systems**

- **Large urban districts and multi-building complexes:** Heat pump systems are ideal for supplying heat and cooling to large urban districts and multiple buildings utilizing a centralized heating and cooling system that significantly curtails energy and operating costs for the entire community.

2.4.3 Variable Refrigerant Flow (VRF) Systems for Energy Savings

The Variable Refrigerant Flow (VRF) systems are considered to be advanced HVAC technologies that would increase energy efficiency, improve thermal comfort, and control the climate in residential, commercial, and industrial buildings effectively. Unlike conventional HVAC systems that use a single-speed compressor and constant flow rates, the VRF systems adjust the refrigerant flow according to real-time heating and cooling requirements on multiple points or zones. Such flexibility greatly lowers the amount of energy used and improves overall system effectiveness.

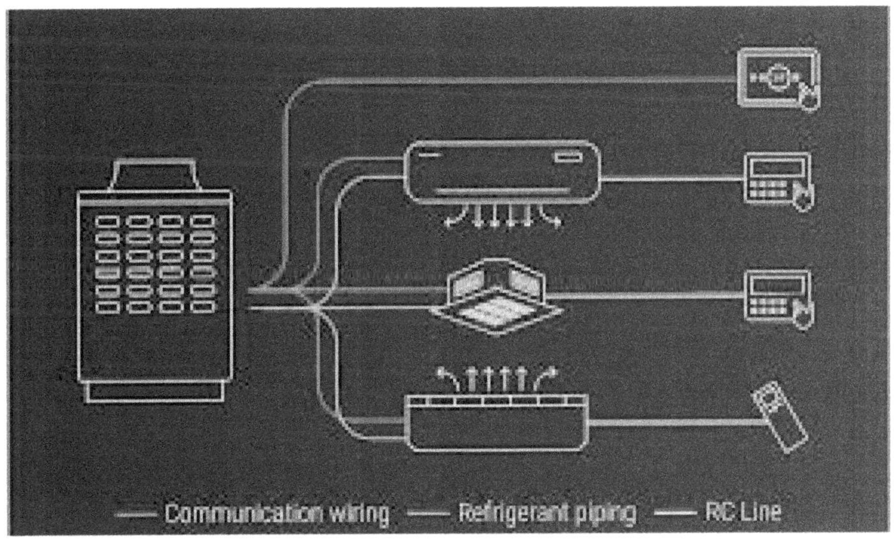

Figure 2.9: Variable Refrigerant Flow (VRF) Systems

Source: (CoolAutomation, 2019)

How VRF Systems Work

The VRF systems are considered to be advanced HVAC technologies that would increase energy efficiency, improve thermal comfort, and control the climate in residential, commercial, and industrial buildings effectively. While conventional HVAC systems employ a single-speed compressor as well as fixed airflow, VRF systems inherently utilize refrigerant flow to match real-time heating and cooling demand on multiple zones. Such flexibility greatly lowers the amount of energy used and improves overall system effectiveness.

Key Components of VRF Systems:

Second is an outdoor Condensing Unit containing an inverter-driven compressor that modulates the refrigerant flow.

- **Indoor Air Handling Units (AHUs):** Capability of independent temperature control in each zone, connected to the refrigerant network.
- **Refrigerant Piping Network:** Transports refrigerant between the outdoor unit and multiple indoor units.

Smart controllers and sensors: From temperature, humidity, and occupancy, we monitor how to operate the system optimally.

Energy-Saving Features of VRF Systems

a. **Inverter-Driven Compressors**

- VRF compressors operate at variable speeds and do not need to turn on and off at full power like traditional HVAC equipment.
- This not only minimizes energy wastage but also the frequency of start-stop cycles, which helps improve the efficiency and durability of the system.

b. **Zoned Temperature Control**

- With different units maintained at different temperatures, VRF systems enable only occupied areas to be heated or cooled.
- It greatly cuts down on energy usage by reducing unnecessary conditioning of unoccupied spaces.

c. **Heat Recovery Technology**

- VRF heat recovery systems transfer waste heat from cooling zones to zones that demand heating, where they are then reconverted to heating utilising the VRF in these buildings that must simultaneously cool as well as heat.
- Plus, this reduces the total energy demand by replacing discarding excess heat with it.

d. **Smart Sensors and AI Integration**

- Featured in modern VRF systems are occupancy sensors, CO_2 detectors, as well as humidity sensors, which adjust airflow and temperature on a real-time basis in line with the environmental conditions.
- AI-driven controllers learn user patterns and adjust settings again, as that would increase system performance.

e. **Reduced Ductwork and Energy Losses**

- Normally, traditional HVAC systems use extensive ductwork work in which there are many energy losses due to wasted air leakage.
- Refrigerant piping rather than ducts is used in VRF systems; there are fewer transmission losses with more improved efficiency.

2.5   Smart HVAC Control Systems and Automation

However, revolutionizing the way HVAC systems are controlled is smart HVAC control and automation systems. The technologies enhance the efficient operation of the buildings, improve indoor comfort, reduce energy consumption, and enhance control of the buildings' climate. Smart systems for HVACs can be learned in real-time through the use of advanced sensors, connections, and machine learning algorithms.

What is Smart HVAC?

A smart HVAC system, which is a logical development from traditional Building Automation Systems (BAS), combines networked HVAC components with Internet of Things technology. The main goal is to give residents the ability to precisely regulate the temperature, lighting, humidity, and fan speed of their rooms. Users may conveniently control HVAC operations using their smartphones or tablets, unlike traditional systems.

Components of a Smart HVAC System

1. The smart HVAC system integrates unique sensors with control boards installed inside individual components. Real-time system data processing, along with algorithm execution, presents itself through components that establish uninterrupted system-to-system communication.

2. The technology deploys control panels throughout different HVAC equipment, while traditional HVAC systems use one central thermostat panel. Through its distributed intelligence framework, the system can detect and modify both incoming and existing house variables.

3. The essential component of Smart HVAC systems is their connectivity infrastructure with Internet of Things (IoT) components. The users get an immediate control of their HVAC system using features such as cellular connectivity, geofencing and voice activation. A wide spectrum of functionalities becomes available through IoT platform integration including sophisticated programming and remote monitoring.

4. The additional intelligent feature known as zoning serves to enhance the functionality of Smart HVAC systems. Through its zoning technology the system gives users complete control to set and adjust temperatures independently within different home areas. The operation remains efficient without zoning because the system runs its cycles for extended periods.(velosiot, 2024).

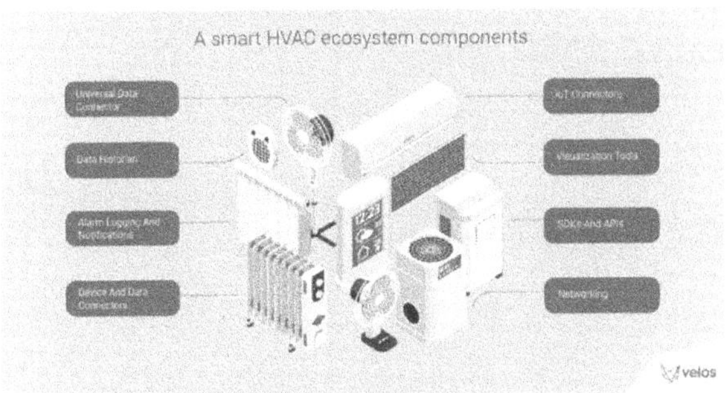

Figure 2.10: A Smart HVAC ecosystem component

Source: (velosiot, 2024)

Benefits of Smart HVAC Control Systems: Energy Efficiency and Cost Savings

- **Real-time Control:** Sensor and algorithm-based real-time control of the system based on real conditions, such as occupancy and ambient temperature alone. As spaces are unoccupied, energy is saved by way of avoiding unnecessary cooling or heating.
- **Advanced systems have Energy Consumption Data:** building managers can track and track inefficient energy usage and optimize the HVAC system to save cost.
- **Energy Demand Shifting:** Smart systems can shift HVAC systems on and off and automate that, as needed, to ensure when there is peak demand on the electrical grid that strain is lessened and energy bills are reduced.

a. **Improved Indoor Comfort**

- **More Precise – Consistent Temperature Control:** This has to do with having available Smart HVAC systems that have precise and consistent temperature control. They are equipped with real-time feedback sensors from sensors to make sure temperatures stay within the building occupants' desired range.

- Smart thermostats allow occupants to personalize their space (humidity, temperature, and air quality) in a manner such that they are comfortable in each of the different zones of a building.

b. **Enhanced Indoor Air Quality (IAQ)**

- Sensors that monitor CO_2 levels trigger an increase in airflow of automatic ventilation systems to maintain a healthy and comfortable environment.
- **Air Filtration and Humidity Control:** Intelligent HVAC systems may be linked with air filtration and humidity control systems for optimal air quality and avoiding the development of mold or allergens.
- **Hands Across a Room:** If contaminants continually enter the room, monitors near the walls can detect their presence and signal the HVAC system to respond accordingly.

c. **Predictive Maintenance and System Longevity**

- **Smart HVAC:** One of the benefits of Smart HVAC is its real-time diagnostics, where the HVAC system continues its monitor system performance and detect early signs of malfunctions or inefficiencies. In so doing, predictions are made of upcoming maintenance and parts replacement, eliminating the need to perform expensive repairs when they fail or when there is downtime.
- **Optimized Operation and Timely Maintenance Alerts:** Through the extended life of HVAC components, their corresponding wear and tear are lowered, enhancing the operational lifetime of the system as well as long-term operational efficiency(Jane Marsh, 2024).

IoT Applications in HVAC Optimization:

- **Smart Thermostats**

Tapping into anal og is something like Nest, Ecobee, and Honeywell that the thermostat can learn user preference over time and adjusts temp settings some connective. Remote mobile apps allow them to be controlled, making them convenient as well as energy efficient. Let's take a look at an example where a smart thermostat can learn the occupancy patterns and optimize the settings according to the user behavior, and ultimately save the heating and cooling costs.

- **Occupancy Sensors**

Occupancy sensors can sense whether a room is in use, and these can adjust HVAC system as per the case avoiding the unnecessary energy use. For example, these sensors can be deployed in a conference room in a commercial building so that air conditioning is only activated when there are meetings and not when the room is left unoccupied.

- **IoT-Enabled Ventilation Systems**

Sensors that measure CO_2 levels, humidity, and pollutants, among other things, monitor indoor air quality (IAQ) inside IoT-enabled ventilation systems. Ventilation rates of this system are adjusted to optimum air quality. Thus, for example, sensors can be used by a hospital to ensure clean air in operating rooms to maintain energy efficiency and patient safety.

- **Introduction to Building Automation Systems (BAS)**

A BAS is a system that controls heating, ventilating, air conditioning (HVAC), lights, and other building systems in a similar way to access control systems. It can also manage escalators, elevators, and irrigation systems, and may share software and hardware with access control systems.

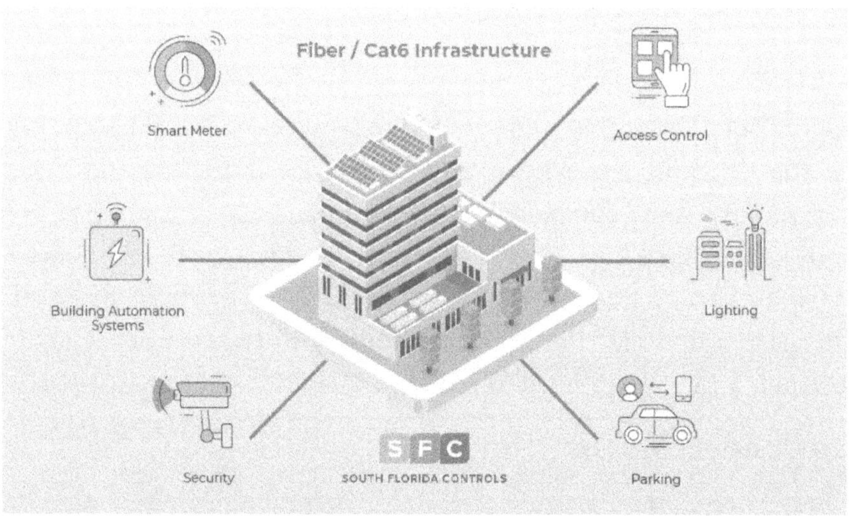

Figure 2.11: Building Automation Systems (BAS)

Source: (Innovations, 2021)

BAS are central systems that manage and automate various building systems such as: HVAC, refrigeration, plumbing, electrical power, lighting, fire protection, life safety, and other building systems. The integration of these building systems through BAS allows for automated control, management, and real-time monitoring, enabling greater operational efficiency as well as improved occupancy comfort. The automation of the operation of multiple building systems under BAS allowed for achieving complex functionalities and sequences of operation that were not possible before older pneumatic control systems, for example. (Ehrlich, 2018).

How Building Automation Systems (BAS) Work

The BAS brings different building systems, like HVAC, into a single platform. With an interface that allows operators to monitor and control the varying areas of a building's environment in one place, this consolidated system gives them one view of everything.

This is achieved by BAS operation in harmony with sensor, controller, user operated devices, and actuator components. All together, these make data collection, make real-time adjustments, and maintain the building in optimal performance utilizing resources.

To be a successful BAS, the data gathering should involve sensors from across the building. These sensors monitor different variables such as temperature, humidity, and occupancy depending on which BAS was designed for that building.

This data is then processed into information by controllers that decide when to adjust the building systems. These commands act as input and as an input, which means that actuators and valves turn things on or off, increase or decrease airflow, adjust water flow, etc. With the system's hostile real-time feedback loop, you can make continual (and continual mistakes) adjustments, ensuring efficiency and comfort. (Abiz, 2024).

2.5.1 IoT-Enabled HVAC Optimization

HVAC systems have become 'smart', 'connected' things, where the IoT has enabled the transformation of traditional, manually controlled systems into real-time performance-optimizing ones. IoT-enabled HVAC systems embed sensors, controllers, or data analytics in HVAC components to provide intelligent, data-guiding approaches towards improving energy efficiency (cost), reducing operating cost, enhancing comfort, and monitoring HVAC system health.

The addition of IoT with HVAC systems is one of the main drivers of intelligent buildings, which provides more control over the environment and less energy loss.

IoT Applications in HVAC

- **Smart Thermostats:** One of the most popular IoT-enabled devices in HVAC systems is, of course, smart thermostats. Based on these devices, learning them, you can reduce your energy consumption and even keep these devices to adjust the temperature automatically to match your schedule. Examples that are popular include the Nest Learning Thermostat and EcoBee Smart Thermostat.

- **Occupancy Sensors:** Occupancy sensors will be used to sense whether a room is occupied or not and adjust the HVAC system accordingly. This includes reducing the air conditioning or heating in empty parts of a commercial building without affecting occupants in other parts of the building.

- **Air Quality Monitoring:** Air quality sensors are smart, and they monitor pollutants, allergens, and other harmful particles in the air. The use of these sensors can activate the HVAC system to run the air filtration or switch ventilation rates to improve the indoor air quality and occupant comfort.

- **Chiller Optimization:** In areas such as large commercial or industrial buildings, chillers are generally the most energy-intensive HVAC component. According to Io Sentinel, IoT sensors and data analytics can also be used to improve chiller operation by making them run at the peak efficiency level, possibly saving energy and thereby lowering wear and tear on equipment.

- **Demand-Controlled Ventilation:** IoT based HVAC systems can be responsive to occupancy, air quality, environmental conditions etc. and can adjust ventilation level accordingly in real time. For instance, the system can increase ventilation to office building during its peak hours of human occupation and decrease for periods when fewer people are present, achieving energy efficiency as well as better indoor air quality.

Challenges of IoT in HVAC

While IoT brings numerous benefits to HVAC systems, it also presents certain challenges:

- **Security Risks:** When IoT systems connect to the Internet, they also become hackers. HVAC systems can potentially be compromised by hackers to their performance or damage. To protect the IoT-connected HVAC systems from security breaches, it is imperative to have strong cybersecurity protocols and encryption methods.

- **Data Overload:** As the IoT technology becomes popular, data generated by these systems is overwhelming to some organizations that are managing such data. Before making any inferences from the data, it is necessary to have good data management and analytics tools to avoid information overload.

- **High Initial Costs:** You can assume that implementing an IoT based HVAC system can be quite expensive on the project front with purchasing sensors, actuators and other smart devices as well as installation and integration with our existing systems. But, although, these costs are compensated by long-term energy savings and efficiency.

- **Interoperability Issues:** That said, not all IoT devices are compatible with each other, meaning that when integrating in HVAC systems or any other building management technologies, this could create problems along the way. The most important factors of making the most out of IoT in HVAC lie in ensuring compatibility and performance integration.

Future Trends in IoT-Enabled HVAC Systems

The future of IoT in HVAC is very promising, as the technology evolves with time. Some of the key trends we should get used to include:

- **Edge Computing:** Using edge computing, IoT devices will have faster access to as well as a capability to process data locally, without reducing the latency and for increasing efficiency. This would be faster to make decisions and reduce reliance on cloud-based platforms.

- **Blockchain for Security:** Blockchain around IoT-enabled HVAC systems can improve the security aspect by ensuring data integrity and safeguarding against cyber-attacks.

- **Integration with Smart Grids:** IoT-enabled HVAC systems will gradually integrate with smart grids to permit buildings to communicate with local energy grids and utilize grid energy to conserve energy through real-time demand(Manvendra Kunwar, 2024).

2.5.2 AI and Machine Learning for Predictive HVAC Management

Artificial Intelligence (AI) and Machine Learning (ML) integration in HVAC have changed the way buildings manage their heating, cooling, and ventilation. With the use of high-end data analytics and predictive algorithms, AI and ML allow the HVAC systems to predict and respond to environmental conditions, occupancy patterns, and HVAC system performance in an instant. Optimum energy consumption and, hence, reduced operation costs and improved occupant comfort are realized.

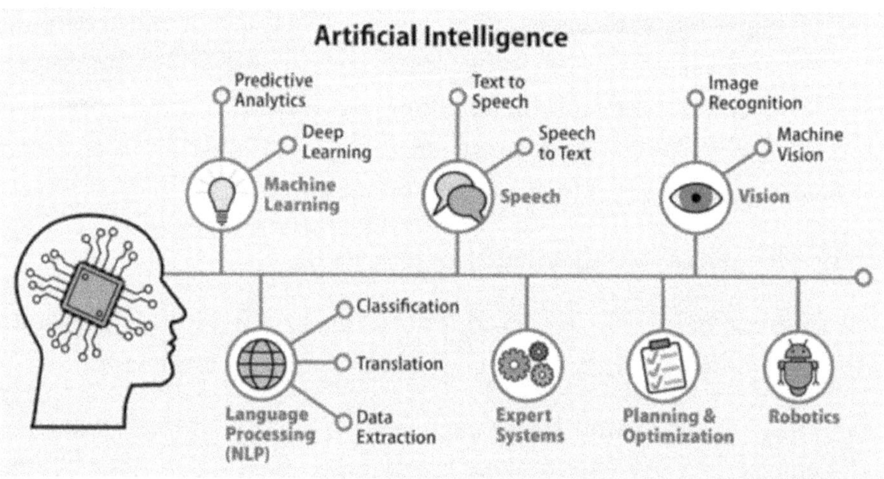

Figure 2.12: Artificial Intelligence (AI),

Source: (Vikram Murthy, 2024)

These technologies have enabled HVAC systems to move beyond traditional reactive management, specifically predictive maintenance, dynamic performance optimization, as well as learning from past data towards further improvement of future system operations. The shift to predictive HVAC management is based on data that gives actionable insights to the building owners and facilities managers so that building performance, the energy efficiency of systems, and sustainability become radically better.

The Future of AI in HVAC Systems

And in the future, there will be more advanced and improved HVAC systems thanks to advanced AI technology. Here are some potential developments:

- **AI-Driven Energy Grids:** The future HVAC systems can be linked with the AI-driven energy grid to support in energy use optimization depending on the grid's demand and availability. Greater energy efficiency together with reduced costs will result from this improvement.

- **Machine Learning for Enhanced Diagnostics:** HVAC systems may be able to self-diagnose problems more quickly and accurately thanks to advanced machine learning algorithms, which would lower maintenance costs and downtime.

- **Integration with IoT:** The future HVAC systems will rely heavily on IoT integration, which refers to the IoT. AI-powered HVAC systems will connect to various building sensors through automation to establish entire smart homes or building platforms. (HVAC Automation, 2024).

2.6   Chapter Synthesis

The exploration of this chapter focused on advanced HVAC technologies, along with Building Automation Systems (BAS), together with IoT and AI integration in building management operations. Real-time sensors and smart control in such systems further increase energy efficiency while reducing costs with comfort and air quality. BAS provides centralized system control, which delivers better operational results. Even with the barriers of security and high costs, IoT-based HVAC optimization can be made automated based on occupancy and environmental conditions. Advances in AI and machine learning, predictive maintenance, and dynamic energy optimization have made the application of HVAC performance further. Put together, these technologies are remodeling building management into a more effective, integrated, and attractive system.

Multiple Choice Questions (MCQs)

1. **Which of the following is essential for designing energy-efficient HVAC systems?**

   a. Ignoring building orientation
   b. Focusing only on heating equipment
   c. Considering insulation and building envelope
   d. Relying solely on traditional energy sources

2. **What is a primary factor in optimizing heating and cooling efficiency in HVAC systems?**

   a. Using low-efficiency boilers
   b. Utilizing advanced heat pump technologies
   c. Avoiding thermal comfort considerations
   d. Limiting air quality control measures

3. **Which HVAC system is known for providing energy savings by adjusting refrigerant flow according to demand?**

   a. High-efficiency boilers
   b. Variable Refrigerant Flow (VRF) systems
   c. Air-to-air heat pumps
   d. Radiant heating systems

4. **In the context of HVAC, what does IAQ stand for?**

   a. Indoor Air Quality
   b. Indoor Atmospheric Quality
   c. Industrial Air Quality
   d. Intelligent Air Quality

5. **Which of the following is a significant factor in system design to achieve energy efficiency in HVAC?**

   a. Poor insulation
   b. Advanced ventilation systems
   c. Ignoring thermal comfort
   d. Over-sized equipment

6. **How can IoT contribute to HVAC energy efficiency?**

   a. By reducing the need for HVAC maintenance
   b. By enabling real-time data collection for optimization
   c. By increasing the power consumption of the system
   d. By reducing the quality of air circulation

7. **Which technology is used to improve HVAC system performance by predicting demand and adjusting operations automatically?**

   a. AI and Machine Learning
   b. Solar panels
   c. Manual thermostats
   d. Steam heating systems

8. **Building automation systems (BAS) are used in HVAC systems to:**

   a. Automate lighting and security only
   b. Control and optimize HVAC functions based on real-time data
   c. Replace traditional HVAC equipment
   d. Perform only cooling functions

9. **What is one of the best practices for ensuring an HVAC system's energy efficiency?**

   a. Using oversized systems for higher capacity
   b. Ignoring the need for regular system maintenance
   c. Integrating high-efficiency boilers and furnaces
   d. Avoiding energy-saving technologies to reduce initial costs

10. **Why is the building envelope an important factor in HVAC system energy efficiency?**

    a. It helps in increasing the external temperature of the building
    b. It prevents heat transfer, helping to maintain stable indoor temperatures
    c. It requires constant modification to improve energy use
    d. It increases the amount of energy consumed for ventilation

Answers

1	2	3	4	5	6	7	8	9	10
c	b	b	a	b	b	a	b	c	b

CHAPTER 03

RENEWABLE AND SUSTAINABLE HVAC SOLUTIONS

3.1 Chapter Overview

This chapter introduces the study's four main themes: integrating heat recovery systems, "Phase change materials" (PCM) and "Thermal Energy Storage" (TES) into "Heating, Ventilation, and Air Conditioning" (HVAC). It goes into detail about how these technologies can lower energy use and promote sustainable development, particularly when it comes to commercial and industrial establishments. The chapter discusses the operation of the TES and PCM as thermal energy storage and release means and their contribution to reducing peak energy demand and thermal regulation, respectively. The importance of heat recovery methods such as "Energy Recovery Ventilation" (ERV) and "Heat Recovery Ventilation" (HRV) in utilizing waste Heat to lower energy consumption is also covered. It summarizes the advantages of putting these systems in place to save operating expenses and carbon footprints, as well as the difficulties involved. The main scope of this chapter is to provide the basis for further economic and environmental analysis of integrating the aforementioned technologies into HVAC systems. Additionally, it underscores the crying need for innovation and development in the field, specifically in easing the adoption as well as in exploiting possible energy optimization. This serves as the basis of understanding how these technologies further sustainable building practices and minimize the environmental effects of energy consumption.

3.2 Integrating Renewable Energy with HVAC

Renewable energy sources are increasingly being included in HVAC systems due to the desire to reduce energy consumption, carbon footprints, and make HVAC systems more sustainable. Because conventional HVAC systems generally rely on nonrenewable power sources to generate electricity, they contribute to both energy consumption and Greenhouse Gas (GHG) emissions. However, using "Renewable Energy Sources" (RES)is a sustainable and environmentally friendly way to address these problems. This section covers the different kinds of renewable energy sources, how to incorporate them into the HVAC system, and how to improve energy efficiency.

Figure 3.1: Renewable Energy with HVAC

Source: (Sean McCleland, 2024)

Benefits of Integrating Renewable Energy in HVAC Systems

1. **Reduced Carbon Footprint:** By using renewable energy (RE) instead of fossil fuels, HVAC systems can drastically lower their carbon impact. This move encourages cleaner air and a healthier environment in addition to assisting in the mitigation of climate change.

2. **Enhanced Energy Efficiency:** A reliable and effective energy source is provided by RES, especially solar and wind power. Energy efficiency is ensured by integrating these sources into HVAC systems, which lowers overall consumption and energy costs.

3. **Sustainability and Long-Term Savings:** By making use of limitless natural resources, HVAC system investments in renewable energy encourage sustainability. As renewable energy becomes more reasonably priced in comparison to conventional energy sources, this investment eventually results in significant cost savings.

Types of Renewable Energy Sources for HVAC Systems

HVAC systems can, however, be powered by or supplemented by a variety of additional RES. But each one provides its unique strengths, and the need of the building in terms of energy requires the respective need of energy source.

1. **Solar energy:** Sunlight is one of the fastest and most widely distributed energy sources on Earth. The amount of solar energy reaching Earth's surface every hour exceeds the planet's annual energy needs. Solar power seems to be the most promising RES, yet our actual usage of it depends on factors such as:

- The hour of day
- What time of year is it
- The exact location

Figure 3.2: Solar energy

Source: (EDF, 2024)

2. **Wind energy:** One clean and abundant energy source is wind. Particularly in the UK. Since wind power currently accounts for about 29.4% of the country's electricity, wind farms are becoming more common throughout the UK.

Figure 3.3: Wind energy

Source: (EDF, 2024)

Wind turbines come in two primary varieties: onshore and offshore. Although offshore wind generates more electricity than onshore options, its initial costs are significantly higher due to its more complex construction... As a result, both are crucial to the renewable energy supply and the power market.

3. **Hydro energy:** Hydropower is a widely used renewable energy source. The construction of a dam or other barrier can be used to harness the power of a large reservoir's controlled water flow, which in turn powers a turbine and generates electricity. This energy is Dependable

- Simple to store
- It is less expensive to set up than other renewable sources.

The ability of hydro to deliver a reliable energy supply is essential in the field of renewable energy. Even though wind and solar power are excellent, it is essential to have alternative renewable energy sources to support them when they are unable to produce electricity.

Figure 3.4: Hydro energy

Source: (EDF, 2024)

4. **Tidal energy:** This type of hydropower powers turbine generators by harnessing twice-daily tidal currents. Although tidal flow is not as reliable as some other hydro energy sources, it is very predictable and may make up for times when the tide is low. A recent survey found that the UK is sixth in the world for tidal energy scientific accomplishments, even though it is occasionally less well-known than some of the other categories on the list.

Figure 3.5: Tidal energy

Source: (EDF, 2024)

5. **Geothermal energy:** Geothermal energy is a great way to heat your home because it uses the earth's inherent heat to create power. Although it produces electricity right under our feet, the United Kingdom uses a small amount of geothermal energy compared to nations like Iceland, where geothermal heat is abundant.

Figure 3.6: Geothermal energy

Source: (EDF, 2024)

6. Biomass energy is the method by which solid fuels produced by plants are converted into electrical power. Biomass generation has always included burning organic materials to create electricity, but newer, cleaner processes are much more efficient. By transforming municipal, agricultural, and residential waste into solid, liquid, and gas fuel, biomass generates energy at a far lower environmental and economic cost.

Figure 3.7: Biomass energy

Source: (EDF, 2024)

Future Trends in Renewable Energy and HVAC Integration

Renewable energy sources and HVAC systems can work together in the future. Continued advancement in technology and continued decrease in costs of renewable energy solutions are expected to make renewable energy solutions available, more accessible, and more efficient. Future trends include:

- **AI, machine learning, and smart HVAC systems:** As smart HVAC systems have become more prevalent, which will be outfitted with sensors, AI, and machine learning, renewable energy sources will be better integrated through waste reduction and energy efficiency.

- **Increased HVAC System Reliability and Efficiency:** As energy storage technologies advance, many buildings will be better able to store extra renewable energy for use during off-peak hours, eliminating the need for backup generators. resulting in less need for operation of the HVAC system during times when it is not needed.

- **Decentralized Energy Grids:** Renewable energy-powered HVAC systems will be easier to utilize as decentralized energy grids (where multiple buildings or communities generate their power) grow.

- **Government Incentives:** To promote the use of RES and energy-efficient HVAC systems, governments worldwide are offering an increasing number of subsidies and incentives. This will certainly serve as an incentive to drive growth in this sector.

3.2.1 Solar-Assisted HVAC Systems

Utilizing solar energy to power HVAC systems is a new frontier in the quest for greener, more efficient technologies. These systems use solar energy for heating and cooling, which offers several advantages for the environment, energy use, and overall financial savings. This in-depth study will examine the operation of solar-powered HVAC systems, their efficacy in diverse settings, and their effects on important environmental variables.

Figure 3.8: Solar-Assisted HVAC Systems

Source: (REvolution, 2020)

Overview of Solar-Assisted HVAC Systems

Sun-assisted HVAC systems convert solar energy into usable power for cooling and heating. Solar energy can be utilized to offer thermal energy to support the system's heating and cooling functions or electricity to power HVAC components.

The types of solar-assisted HVAC systems are of two major types:

- Solar Photovoltaic (PV) HVAC Systems – A Conventional HVAC unit is powered using electricity generated by the use of solar panels.
- Solar thermal HVAC systems employ solar energy captured in collectors to heat water, cool spaces, or both.

Benefits of Solar-Assisted HVAC Systems

1. **Energy Cost Savings:** Solar-assisted HVAC systems reduce energy bills and are good for their financial bottom line by reducing reliance on grid electricity or fossil fuels. Incentives, rebates, and tax credits are provided by many governments to add to the lower cost-effectiveness.

2. **Environmental Sustainability:** Solar-assisted HVAC systems reduce greenhouse gases by disposing of nonrenewable fossil fuels and using renewable solar-powered energy. This also helps to meet sustainability goals and reduce rates of climate change.

3. **Increased Energy Independence:** Solar-assisted HVAC lets businesses and homeowners generate power on their own, lowering the dependence on utility providers and energy price fluctuation.

4. **Improved HVAC Efficiency:** Solar thermal or solar-assisted heat pump systems work in conjunction with HVAC units to deliver these parts of heating or cooling needs, thus reducing strain on the system and extending HVAC equipment's lifespan(VR, 2023).

Solar-powered HVAC systems come in a variety of designs, including:

1. **Systems that use solar heat:** These systems employ sun radiation to warm a fluid, which is subsequently used to generate hot water for the building or to heat the space.

2. **Photovoltaic (PV) systems:** These systems employ solar cells to convert sunlight directly into energy, which powers HVAC equipment.

3. **Hybrid systems:** These systems combine PV and solar thermal technologies to give the building ventilation and warmth.

All things considered, solar-powered HVAC systems offer a cost-effective and sustainable way to power a building's HVAC system.

A structure's success may depend on its location, temperature, and energy needs as well as the availability of solar resources.

Challenges and Considerations

- **Initial Investment**: High installation cost, but long-term savings and incentives lower the expense.
- **Solar Supply:** supplies energy to be used intermittently and requires storage or backup energy sources.

- **Space Requirements**: Solar panels or solar collectors require sufficient roofing or ground space.

Integration with existing HVAC infrastructure may require upgrades and compatibility with it.

3.2.2 Geothermal Heat Pumps: Design and Efficiency Considerations

Geothermal heat pumps are machines that extract heat from the earth's surface, underground, or both. These heat pumps can be found underground and are also called geo-exchange. These systems are referred to as ground-coupled heat pumps, surface water heat pumps, and groundwater heat pumps, respectively, based on their configuration. The Commonwealth Building, which was built in 1946 in Portland, Oregon, was the first commercial project to be completed successfully. There were 12 gigawatts of geothermal heat pump thermal capacity installed in the US as of 2004, and 80,000 more were being added annually.

Comparing geothermal heat pumps to traditional heating or cooling systems, the former consumes 25% to 50% less electricity. They require less maintenance, are quieter, last longer, and are independent of the ambient air temperature than air-source heat pumps. The decision to use this technology may be influenced by utility pricing for natural gas, electricity, or other fuels.

Although the majority of locations in the US can use geothermal heat pump technology, specific site features will determine the kind of system that works best for a given location. Access to open water sources, local groundwater availability and temperatures, available ground area, and the surrounding soil's thermal conductivity can all further guide their utilization in a project. (wbdg, 2023).

Types of Geothermal Heat Pump Systems

There are four main types of GHP ground loop systems. The vertical, horizontal, and pond/lake systems are all designed to be closed-loop. The fourth type of system is the open-loop option. Some factors, including the site's temperature, soil type, available land, and installation costs, determine which option is ideal. These methods are versatile enough to be used in both residential and commercial construction. By assessing the site's soil and ground composition and talking about the intended use, an accredited contractor or installer can decide which kind of system is ideal to install in a given area (Energysaver, 2023).

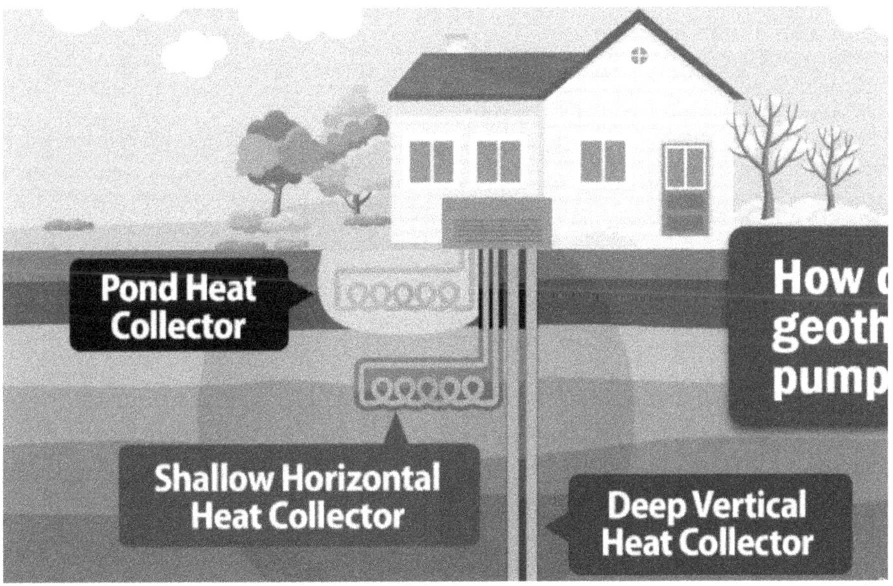

Figure 3.9: Geothermal Heat Pump Systems

Source: (energysaver, 2023)

1. **Closed-Loop Systems:** Most closed-loop geothermal heat pumps use a closed loop with polyethylene tubing which is buried in ground or submerged in water and circulates water or a water and

glycol mixture. The heat pump's refrigerant and the closed-loop antifreeze solution transfer heat through a heat exchanger.

One kind of closed-loop system that essentially pumps the refrigerant through horizontal or vertical copper tubing that is buried in the ground is a direct exchange, instead of a heat exchanger.

2. **Horizontal:** One pipe is often buried six feet, and the other four feet in a two-foot-wide trench, or two pipes spaced five feet apart necessitate ditches that are four feet deep. In most cases, residential installations are the most cost-effective for this type of installation, particularly in cases when available acreage is sufficient for new construction.

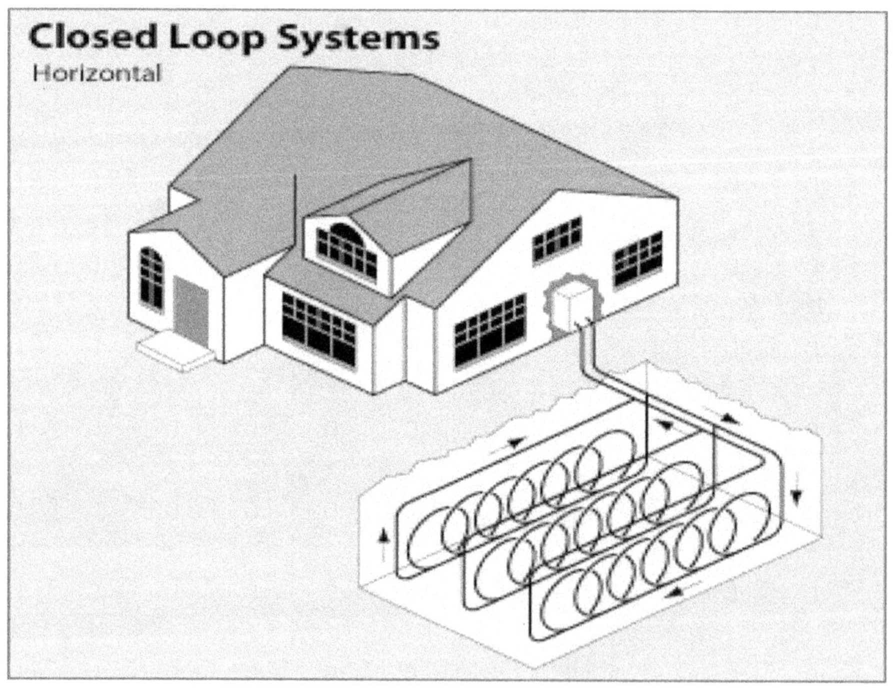

Figure 3.10: Horizontal Closed Loop Geothermal System.

Source: (energysaver, 2023)

3. **Vertical:** The process involves boring holes 100 to 400 feet deep, 20 feet apart, and about four inches in diameter. Two pipes can then be placed and joined at the base using a U-bend to form a loop. Large commercial structures and educational institutions often adopt vertical systems since the land area required for horizontal loops is too high.

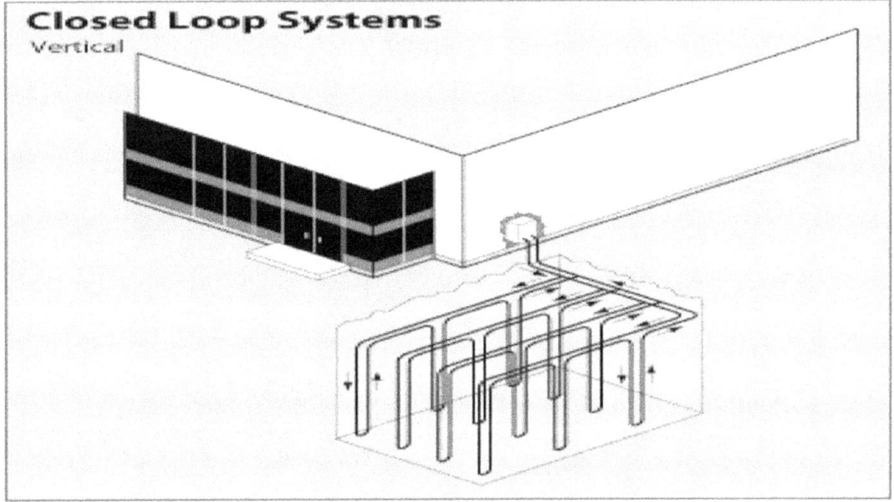

Figure 3.11: Vertical Closed Loop Geothermal System.

Source: (energysaver, 2023)

4. **Pond/Lake:** Geothermal heat pumps, in contrast to conventional heat pumps, use an artificial body of water that meets specific depth, volume, and quality requirements to transfer heat. Running helically from the building to the water is an underground supply line pipe. To avoid freezing in colder climates, these rings are positioned at least eight feet below the surface.

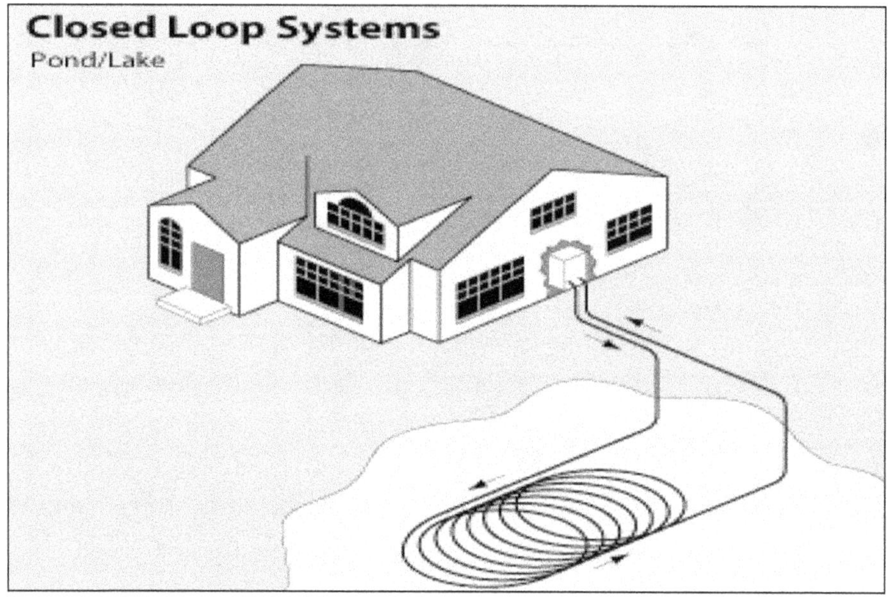

Figure 3.12: Pond/Lake Closed Loop Geothermal System.

Source: (energysaver, 2023)

5. **Hybrid Systems:** These systems use many geothermal resources or combine one with external air (like a cooling tower). When cooling needs are much greater than heating requirements, hybrid techniques shine.

Energy-Efficiency Advantages of "Geothermal Heat Pumps" (GHP)

"Geothermal heat pumps" (GHP) are hard to beat when it comes to energy efficiency. These systems can be up to 45% more efficient than traditional HVAC systems since they rely on the consistent temperature below ground instead of the fluctuating air temperatures. This means they can consistently provide heating and cooling without consuming excessive amounts of energy.

Geothermal systems are the most cost-effective, environmentally responsible, and energy-efficient HVAC option available, based on the

U.S. Environmental Protection Agency (EPA). Homeowners who set up geothermal systems can expect to see significant reductions in their energy bills, often cutting heating and cooling costs by as much as 70%.

One of the primary reasons geothermal heat pumps are so efficient is that they don't rely on combustion to generate heat, unlike gas furnaces or propane furnaces. Instead, they transfer heat, which requires much less energy. This is an environmentally favorable choice since it lowers energy use and GHG emissions. (Mike Haine, 2023).

3.2.3   Hybrid HVAC Systems and Alternative Energy Sources

A gas furnace and an electric heat pump make up a hybrid system, also referred to as a dual-fuel heat pump. Depending on the season, temperature, and function needed, the system alternates between using each of the two units to heat and cool your home as efficiently as possible throughout the year.

During the summer, heat from the system pump functions similarly to a central air conditioner by moving hot air outside your house until your thermostat registers the temperature you want. By supplying economical heat during the milder months, In the autumn and spring, the heat pump also does the majority of the work.

When the temperature drops during the chilly winter months, the furnace steps in and works to heat your entire house. A furnace's primary job is to swiftly and effectively transform fuel into heat. Your hybrid heat system's furnace will do just that when it's cold outdoors.(Trance, 2023).

What is a hybrid system in HVAC, and how does it work?

In a dual-fuel system, a hybrid HVAC system (sometimes an electric heat pump and a gas furnace) is created. A dual-fuel heat pump system of this type is a variation of this type of technology. Dual fuel is gas, propane, or oil for the furnace and electricity for a heat pump. The hybrid system is also packaged.

The air source heat pump is the main source of heating and cooling for your house from spring through fall. The furnace acts as an air handler, delivering warm air or cool air to your home.

Advantages of a hybrid HVAC system

Enjoy year-round comfort with a heat pump and wintertime heating and cooling efficiency with a furnace.

1. **Energy efficiency:** Heat pumps that use air sources can have an efficiency of over 100%. This indicates that they can generate 300% more thermal energy than they use for electricity. In general, heat pumps are more effective at heating than furnaces (in the proper temperature range) and more efficient than central air conditioning devices.

2. **Cost savings:** The Department of Energy (DOE) estimates that replacing an air conditioner with a heat pump can result in annual savings of over $500. Additionally, you can save anywhere from 30 to 50 percent of your overall heating and cooling expenses by combining that heat pump with a gas furnace to create a hybrid system!

3. **Less carbon footprint:** Heat pumps lessen your carbon footprint by heating your house without using fossil fuels like oil, natural gas, or propane. For this reason, there are numerous financial incentives available to American households to encourage them to switch to heat pumps for heating and cooling, including heat pump tax credits and heat pump rebates.

- **Alternative Sources of Energy**

The term "alternative energy" refers to the generation of power from sources other than the more common and environmentally harmful fossil fuels like coal, oil, and natural gas. Included are both renewable and nuclear power.

The following are the most widely utilized alternative energy sources:

1. Wind Energy
2. Solar Energy
3. Geothermal Energy
4. Bioenergy Hydroelectric
5. Energy Hydrogen
6. Nuclear Energy
7. Tidal Energy

Below is an explanation of the energy sources mentioned above:

Applications of Alternative Energy Sources

1. **Wind Energy:** This energy source uses wind to create electricity by harnessing the kinetic energy of flowing air.

Figure 3.13: Wind Energy

Source: (Nafeez khan, 2023)

2. **Solar Energy:** This energy source makes use of solar radiation, which has the power to generate heat, chemical reactions, and electricity.

Figure 3.14: Solar Energy

Source: (Nafeez khan, 2023)

3. **Geothermal Energy:** This is power that has been extracted from deep inside the planet. This heat was generated when the planet was first being formed.

Figure 3.15: Geothermal Energy

Source: (Nafeez khan, 2023)

4. **Bioenergy:** This energy source uses biomass, or organic materials, to produce gas and power. One resource that can assist in meeting the energy demand is bioenergy, which includes fuel for vehicles, heat, and electricity.

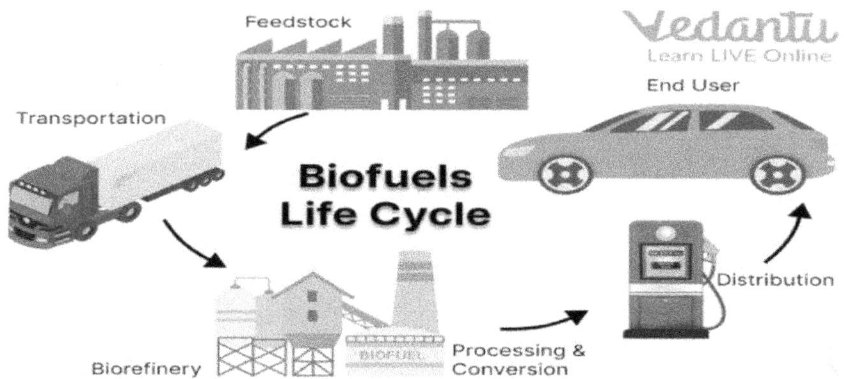

Figure 3.16: Bioenergy

Source: (Nafeez khan, 2023)

5. **Hydroelectric Energy:** Since potential is defined as the ability to flow downward from a specific height, it is a source of energy that is obtained by flowing water from high potential to low potential. A waterfall is an example of a natural feature that may be used to generate electricity through the use of hydropower.

The Value of Alternative Energy Sources

Below is a discussion of Alternative energy sources

- **Preserves the Environment:** Using alternative energy sources promotes environmental preservation and regeneration.
- **Aids in the Creation of Sustainable Fuel Systems:** Alternative energy sources can support the ecological balance of an area and aid in the development of a sustainable fuel system.

- **Aids in Lessening Dependency on Imported Fuels:** Using alternative energy sources also aids in lessening reliance on imported fuels.
- **Contributes to Raising Income:** By giving the populace of the nation more work options, it contributes to raising the nation's income.
- **Beneficial for Fossil Fuel Conservation:** Using alternative energy sources aids in the preservation of fossil fuels.
- **Helpful in Delaying and Reversing Climate Change:** Due to their significantly reduced carbon content, alternative energy sources aid in delaying and reversing climate change.
- **Beneficial to Economic Growth:** Increasing utility-scale energy system production can help the economy grow even more (Nafeez khan, 2023).

3.3   Sustainable Refrigerants and Green Cooling Technologies

Sustainable refrigeration

The F-gas Regulation is a major factor in the industry's overall shift towards more environmentally friendly cooling methods. The law prohibits the use of several F gases when they are more ecologically utilized in refrigeration and air conditioning systems. Y friendly substitutes and restricts their use, particularly those frequently utilized in air conditioning and refrigeration systems. All F gases will soon need to be replaced, but the more dangerous ones are being phased out first.

Cooling agents used in HVAC and refrigeration systems that have little environmental impact (especially as far as ODP and GWP) are called sustainable refrigerants. These refrigerants have been designed with energy efficiency, GHG reduction, and adherence to environmental standards like the Kigali Amendment and the Montreal Protocol in mind.

These have played a key role in climate change as well as ozone layer depletion, being traditional refrigerants such as CFCs, HCFCs, and HFCs. Natural refrigerants, hydrofluoro olefins (HFOs), and other eco-friendly solutions are used in sustainable alternatives, such that the carbon footprint can be minimized.(space, 2023).

Types of Sustainable Refrigerants

1. **Natural Refrigerants:** A lot of interest has been shown in natural refrigerants as safe alternatives to conventional ones. They stand out because of their low potential for global warming, lack of ozone-depleting properties, and negligible or non-existent environmental impact. The three most widely used natural refrigerants are carbon dioxide (CO_2) 25, ammonia (NH_3), and hydrocarbons (HCs).

2. **Carbon Dioxide (CO_2):** At higher pressures, CO_2 refrigeration systems are easier to run, but pose some design and safety concerns compared to conventional systems. But increasingly, the development of technology has rendered CO_2 systems more practicable and efficient, so they can be used in industrial refrigeration, cold storage facilities, and supermarkets. CO_2 is inert since it is non-toxic, low-cost, non-inflammable, and therefore a safe option for refrigeration applications. The CO_2 by-product is another component that maintains the sustainability of the sector, whereby it may be captured and utilized as a refrigerant.

3. **Ammonia (NH_3):** Ammonia (NH_3) is another popular natural refrigerant having impeccable thermodynamic qualities. It is one of the greenest choices because it has a GWP (Global Warming Potential) of zero, and it also has no potential to destroy the ozone layer. In the past, NH_3 has been used for many years in large-scale air conditioning, cold storage, and industrial refrigeration. It is extremely energy efficient, has good heat transmission qualities, and uses very little power. Since it is poisonous, one must handle it very carefully and take proper safety measures so that optimum

performance can be achieved. Therefore, proper system design, along with ventilation and leak detection, is needed to control the dangers of ammonia.

4. **Hydrocarbons (HCs):** Two naturally occurring hydrocarbons (HCs) with a low GHG footprint but a high ozone layer depleting one are propane (R-290) and isobutane (R-600a). Typical HCs apply to commercial refrigeration equipment, vending machines, and modest home refrigeration systems. They are very thermodynamically good and quite energy efficient. Although HCs are effective and have little impact on the environment, they are combustible and require additional safety precautions. To warrant the safe use of hydrocarbon refrigerants, proper system design is essential, including reliable leak detection and ignition prevention methods.

5. **Hydrofluoroolefins or HFOs:** As a less harmful as a possible alternative to conventional refrigerants, a novel family of refrigerants, hydrofluoroolefins (HFOs), has grown in favor. HFOs have a low global warming potential (GWP) and strong thermodynamic characteristics. 1234yf and 1234ze are two commonly used HFOs. 31 The existence of HFOs was a necessity to overcome the problems caused by the traditional refrigerants in terms of their high GWP. Created to slow climate change by reducing GHG emissions, HFOs created a solution with their enthusiasm and problems. Such refrigerants also have a much lower GWP compared to their predecessors, such as hydrofluorocarbons (HFCs). To balance performance and environmental effects, HFOs were created. International rules and programs aimed at reducing GHG emissions are satisfied by HFOs, as the GWP decreases, thus decreasing the refrigeration and air conditioning systems' impact on global warming.

6. **Hydrofluorocarbons (HFCs):** The industry developed hydrofluorocarbons (HFCs) as synthetic refrigerants to substitute hydrochlorofluorocarbons (HCFCs) and

chlorofluorocarbons (CFCs), which depleted ozone. HFCs possess high potential GHG effects, which result in climate change effects despite their lack of ozone layer impact. A search for environmentally friendly alternatives, including refrigerant blends/mixtures with low-GWP values, has taken place to reduce negative impacts. When the market needed sustainable refrigerants, HFC substitutes with low-GWP were developed to meet this requirement. The lower impact on global warming makes these alternatives target the same operational characteristics as HFCs while achieving equivalent thermodynamic properties. Two low-GWP HFC substitutes include HFC-152a with a GWP of 138 and HFC-32 having a GWP value of 675. The substitutes demonstrate substantial GWP reduction compared to common HFCs, HFC-134a (holding a GWP of 1,430) and HFC-410A (GWP of 2,088), while maintaining their operational cooling effectiveness.

7. **Hydrofluoroethers (HFEs):** They are synthetic refrigerants with zero ODP and low to medium GWP. They are not combustible, have excellent material compatibility and thermal stability, and are non-caking. Examples of such refrigerants are R-E347mcc, R-E245fa2, and R-E245cb2. They are mostly used as solvents, blowing agents, and heat transfer fluids also be used as refrigerants for low-temperature applications.

Green Cooling Technologies

Green cooling technologies are environmentally friendly, energy-efficient cooling systems that are intended to reduce air conditioning and refrigeration environmental footprint. The primary objectives of these technologies are to decrease GHG emissions, increase energy efficiency, and utilize sustainable energy sources. They also become particularly important as the environment is under increased stress because of the global increase in cooling demand caused by urbanization and climate change.

Key Features of Green Cooling Technologies

1. **Low-GWP and Non-Toxic Refrigerants:** One aspect of green cooling is the use of green refrigerants, which have zero "Ozone Depletion Potential" (ODP) and low "global warming potential" (GWP). The cooling system's impact on climate change is mitigated by these low GWP natural refrigerants (CO_2, ammonia, and hydrocarbons) and synthetic refrigerants (HFOs).

2. **Energy Efficiency:** Green cooling utilizes energy-efficient systems; such systems consume less power but still maintain good cooling performance. Energy efficiency is enhanced by a variety of technologies, such as smart thermostats, heat exchangers, high-efficiency, and variable-speed compressors.

3. **Renewable Energy Integration:** In order to operate cooling systems sustainably and without depending on fossil fuels, green cooling systems frequently include RES, such as geothermal, wind, and solar.

4. **Environmentally Friendly Design:** Green cooling systems are designed with sustainable materials, modular designs, and maintenance that is easy, all of which support lower environmental impact.

3.3.1 Transition from HFCs to Low-GWP Refrigerants

The transition from HFCs to lower GWP refrigerants is essential to reducing GHG emissions and halting climate change. HFCs are widely used in refrigeration and air conditioning, although they have a very high potential to cause global warming and contribute significantly to it.

The Montreal Protocol and its Kigali Amendment 2016 have defined clear phasing down of HFCs and promotion of low GWP alternatives as part of international agreements. The environmental regulations, technological advances, and spreading industry awareness are driving this transition.

Why Transition from HFCs?

Religious reasons are not driving the movement from hydrofluorocarbons (HFCs) to low GWP (Global Warming Potential) refrigerants, it is all about a reduction in climate impact, compliance with the global regulations for the same, and energy efficiency of heating, ventilation, and air conditioning (HVAC). Due to their widespread use in air conditioning, refrigeration, and heat pumps, HFCs significantly contribute to global warming (RACHP).

- **High Global Warming Potential (GWP) Impact**

The primary reason HFCs are phasing out is largely due to do with their high GWP, which is how a GHG absorbs heat into the atmosphere over a given period (usually 100 years) compared to carbon dioxide (CO_2). With GWPs of more than 1,300 and even more than 3,900, some of the most widely used HFCs (such as R-134a and R-410A) are thousands of times more potent than CO_2 in terms of their ability to cause global warming. Accelerating climate change and global temperature rise, and exacerbating extreme weather and extreme heat and cold conditions, it is essential to move to low GWP refrigerants that create much less harm to the environment.

- **Regulatory Compliance and Global Agreements**

International agreements and national policies also push the transition from HFCs. For instance, the Montreal Protocol, which bans ozone-depleting substances (ODS) like "hydrochlorofluorocarbons" (HCFCs) and "chlorofluorocarbons" (CFCs), has been amended with the Kigali Amendment 2016, which establishes a global phase-down of HFCs. This agreement requires a reduction in HFC production and consumption as well as the use of low-GWP substitutes.

Apart from global treaties, most countries have tightened up environmental laws and carbon reduction goals, which encourage industries to move toward a sustainable cooling solution. This is causing

U.S. The American Innovation and Manufacturing Act (AIM Act), the European Union's F-Gas Regulation, China, Japan, and India should all quickly implement climate-friendly refrigerants and prohibit the use of high-GWP compounds in HVAC and refrigeration systems.

- **Energy Efficiency Improvements**

The switch from high GWP to low GWP refrigerants is due in part to the requirement to adhere to environmental standards – an opportunity exists for improving the energy efficiency of cooling systems. Next-generation refrigerants (HFO, HFC, and natural refrigerants (CO_2, ammonia, and hydrocarbons)) have higher thermodynamic efficiency as compared to traditional HFC counterparts.

This makes it more energy efficient, therefore using less energy, lower operational costs and decreases the carbon emissions associated with the renewable. Since almost 10% of global electricity use goes to HVAC and refrigeration, there is potential to save big amounts of energy through improved efficiency by choosing better refrigerants. In addition, those systems do less strain on power grids during peak demand times in scorching climates.

- **Industry and Consumer Demand for Sustainable Solutions**

There is increasing awareness regarding the need for greener cooling technologies within industries and among consumers alike. A variety of companies in commercial real estate, manufacturing, food retail, healthcare, and automotive sectors are under pressure to integrate low-carbon HVAC solutions with their corporate sustainability goals (ESG).

Customers place a high value on energy-efficient air conditioning and refrigeration systems that use eco-friendly refrigerants. Appliances that not only cut down energy bills but also help minimize the environmental impact are an active search for households and businesses. This has adversely shifted market preference to create an advantage for manufacturers who have invested in R&D of low-GWP HVAC systems and have phased out the use of harmful refrigerants.

Moreover, both industries and consumers are encouraged by financial incentives like government rebates, carbon credits, green certifications, etc., to make a transition. Companies that constructively adopt low-GWP technologies will not only augment their brand recognition but also satisfy corporate social responsibility (CSR) goals and regulatory requirements without major disturbances.

3.3.2   Emerging Eco-Friendly Cooling Technologies

Eco-friendly cooling technologies have made great progress in recent years as a result of the increased need for sustainable and energy-efficient cooling systems. Weaning away from such high Global Warming Potential (GWP) refrigerants, such as HFCs (hydrofluorocarbons), has already been done in traditional HVAC and refrigeration systems as part of the ongoing movement toward minimizing climate change. To address these environmental impacts, researchers and industries have begun to develop next-generation cooling technologies with lower energy consumption, a low environmental footprint, as well as improved operational efficiencies.

This section covers emerging eco-friendly cooling technologies altering the HVAC and refrigeration industry.

1. Magnetic Refrigeration

Magnetic refrigeration is the latest in cooling technology in situations when conventional refrigerants are not required. The phenomenon known as MCE is based on the observation that some materials change temperature when they are subjected to and work with a magnetic field. The principle of operation takes advantage of heat-absorbing and heat-releasing materials that can be magnetically cycled, efficiently cooling.

Key Benefits:

- It eliminates harmful refrigerants (zero GWP, zero ozone depletion potential (ODP)).
- Higher efficiency than conventional vapor-compression refrigeration
- Less maintenance since it has fewer parts moving.

2. Thermoelectric Cooling (Peltier Effect Cooling)

The heat of the Peltier effect is utilized for thermoelectric cooling, where an electrical current flow over two different semiconductor materials generate a temperature difference to cool at one side and heat at the other. Already, this technology is in use in portable refrigerators and electronic cooling systems, and the technology will continue to improve so that it is more practical for use in large-scale HVAC applications.

Key Benefits:

- There was no refrigerant involved; therefore, environmentally friendly
- Lacking moving parts, compact and silent operation
- Can use renewable energy.

3. Evaporative Cooling (Adiabatic Cooling)

Adiabatic cooling refers to the cooling with natural vaporization that occurs without the expense of energy. It evaporates water in the air, that absorbs heat and makes the ambient temperature go down. This technology works excellently, particularly in dry and hot climates with low humidity.

Key Benefits:

- Compared to traditional air conditioning, it is 80% more efficient.
- It is an eco-friendly option without refrigerants required
- Lower operational costs due to reduced electricity consumption

4. Electrocaloric & Baro caloric Cooling, Solid-State Cooling

Other solid-state cooling technology based on advanced materials has the ability to cause a temperature change by application of an electric field or pressure. They function without traditional refrigerant and are candidates for replacement of vapor compression refrigeration.

Key Benefits:

- No harmful gases released
- Higher efficiency than conventional refrigeration systems
- It can be miniaturized for applications with a compact size.

5. Absorption and Adsorption Cooling

In absorption and adsorption cooling systems, heat energy drives the cooling cycle rather than electricity. Ammonia water (absorption) or silica gel water (adsorption) pairs are typically used in these systems to make cooling. Due to their power sources, which are typically waste heat, solar energy, or geothermal energy, they are very sustainable. (dandb, 2023).

Key Benefits:

- The renewable heat sources are utilized, thereby reducing dependency on electricity.
- No synthetic refrigerants required
- Can be used with industrial waste heat recovery systems.

6. Cryogenic Cooling (Liquid Air and Nitrogen-Based Cooling)

They apply cryogenic cooling technologies that use liquefied gases, e.g., nitrogen or air, to cool down specialized applications. These systems operate by storing liquid air at very cold temperatures and expanding it in order to create cold air when necessary.

Key Benefits:

- Also, when the liquefaction sources used are of renewable sources, zero emissions.

- Highly effective for high-temperature industrial processes
- Can be combined with energetic storage systems.

7. Radiative Cooling (Passive Cooling Technology)

Passive cooling technology, known as radiative cooling, would transfer heat to the space without consuming electricity. In fact, special material can both reflect sunlight and radiate infrared heat away from a building or device to lower temperatures below ambient levels, including during the day.

Key Benefits:

- It does not require any energy input. Thereby, it is a completely passive cooling solution.
- It reduces energy costs and also lowers the need for air conditioning.
- Can be integrated into a building material for large-scale cooling.

3.3.3   Regulatory Frameworks for Sustainable Refrigerants

Businesses are being pushed to use sustainable refrigerants by a robust regulatory framework that offers them ways to lessen the environmental impact of cooling systems. Governments, international organizations, and environmental agencies have implemented policies to gradually phase out high-GWP (Global Warming Potential) refrigerants and switch to low-GWP and natural alternatives. The development of renewable HVAC and refrigeration technology is facilitated by actions in this account that lessen climate change, ozone layer depletion, and GHG emissions.

Different countries and regions have different rules, but they all have the same thing in mind: to curtail dependence on HFCs (hydrofluorocarbons) and other damaging refrigerants and promote the use of climate-friendly cooling tools. Among these policies are international and national agreements, industry standards, and incentive programs that shape the market for refrigerants in commercial and residential applications.

1. The Montreal Protocol and Kigali Amendment

A key international agreement concerning refrigerants is the Montreal Protocol on Substances that Deplete the Ozone Layer. This was subsequently approved in 1987 to get rid of ODS, such as HCFCs and CFCs, which are frequently found in air conditioners and refrigerants. Thanks to the success of the Montreal Protocol, CFCs were almost completely eradicated and are helping to recover the ozone layer.

The Kigali Amendment of 2016 builds on this success by tackling HFCs, which, whilst not destroying the ozone layer, have a very substantial potential to cause global warming. This amendment requires developed and developing countries to reduce HFC use in phased fashions. To meet the goal of cutting HFC use by more than 80 percent by 2047, GHG emissions will be reduced, and the global temperature will be limited.

2. The F-Gas Regulation of the European Union

The most stringent application and refrigerant requirements are found in the European Union (EU), namely in the Regulation (EU) No 517/2014, which governs F Gas. Its goal is to gradually replace high-potential greenhouse gases with HFCs for air conditioning, refrigeration, and heat pumping systems through a quota system that regulates the production and sale of HFCs. The main provisions of the F-Gas Regulation are as follows:

- **First, there is a phasedown of HFCs:** The EU aims to reduce HFC use by 79 percent compared to 2015 levels by 2030.
- **Regulation:** As GWP value of the refrigerant is concerned, the regulation limits the use of refrigerants based on their GWP values, thereby stimulating the use of hydrofluoroolefins (HFOs), carbon dioxide (CO_2), ammonia (NH_3), hydrocarbons (propane and isobutane).
- **Prohibition Against Certain Equipment:** New refrigeration and air conditioning equipment are prohibited from using High GWP

refrigerants, forcing manufacturers to come up with sustainable refrigerants.

Many multinational HVAC and refrigeration manufacturers have also adopted these regulations globally, as even if they are not required to do so, they comply with EU standards elsewhere.

3. The U.S. AIM Act and EPA Regulations

The "American Innovation and Manufacturing" (AIM) Act, which was passed in 2020, is the main regulatory framework for lowering HFC emissions in the US. The "Environmental Protection Agency" (EPA) is empowered by this law to take steps to reduce HFC production and consumption. Encourage the use of climate-friendly technologies, and create industry-standard practices for refrigerant management.

Key aspects of the AIM Act:

- **HFC Phasedown:** By 2036, the United States aims to reduce HFC production and consumption by 85%.
- **To encourage research,** development, and the adoption of low GWP refrigerants and sustainable cooling systems, such as cooling towers and thermal storage as well as absorption and absorption heat pipes, the act reduces GHG intensity or accumulates emission units.
- **Reduced Emissions:** Stiffer refrigerant recovery and recycling rule, which limits the release cooling systems.

Furthermore, A major factor in licensing is EPA's "Significant New Alternatives" Policy (SNAP) program. eco-friendly substitutes for use in heat pumps, air conditioning, and refrigeration systems.

4. Regulations in Asia-Pacific and Other Regions

Aggressive efforts are made by many countries in the Asia Pacific region to move to sustainable refrigerants aligned with global needs. They have each imposed their regulatory framework for HVAC and refrigeration in China, Japan, and India, some of the world's biggest HVAC and refrigeration markets.

- Kigali Amendment commitments of China: China is phasing down HFC production and ramping up natural refrigerant investment in CO_2 and other ammonia.
- The Fluorocarbon Law in Japan mandates leak detection and recovery of high GWP refrigerants for proper disposal and incentives for low GWP technologies.
- The India Cooling Action Plan (ICAP) aims to reduce HFC emissions, promote efficient cooling and expand use of the low GWP refrigerants such as hydrocarbons and CO_2.

Further, countries like Australia, Canada, and South Korea have also imposed regulations to cut HFC dependency and encourage cooling technologies.

5. Industry Standards and Certifications

Refrigerants are subject to standard developed by industry organizations in addition to government regulations so that they are safe, efficient and in compliance with their environmental impact. Some notable frameworks include:

- **The American Society of Heating, Refrigerating, and Air-Conditioning Engineers' (ASHRAE) standards:** Define the classification, safety, and performance standards for alternative refrigerants.
- **ISO 5149 (International Organization for Standardization):** Establishes global guidelines for safe handling and application of refrigerants.
- **Green Building Certifications (LEED, BREEAM, WELL):** Promote the use of ecologically friendly and energy-efficient HVAC systems that employ sustainable refrigerants.

The government policies are complemented by these industry standards, which drive the transition to the use of climate-friendly cooling solutions.

6. Financial Incentives and Market-Based Approaches

To enable faster adoption of sustainable refrigerants, most governments and organizations also reward businesses and consumers, who adopt low-GWP cooling systems with financial incentives and carbon pricing mechanisms.

- **Carbon Pricing and Emission Trading Systems (CTS):** These include France, Canada, and the EU, which implement carbon taxes or programming, which encourages businesses to invest in sustainable refrigerants.
- **Tax Credits and Subsidies:** In some cases, some governments give tax rebates and subsidies for installing low GWP HVAC systems, usually for commercial and industrial applications.
- Funding for research and development of next-generation refrigerants to allow manufacturers to innovate and commercialize eco-friendly cooling technologies.

The market-based approaches encourage firms to migrate away from harmful refrigerants and, at the same time, meet the regulatory requirements.

3.4   Thermal Energy Storage and Waste Heat Recovery

However, to meet the demands of sustainability and energy efficiency, "thermal energy storage" (TES) and "waste heat recovery" (WHR) techno WHR and TES logiest are significantly reducing energy consumption in buildings and enterprises. Efficiency, energy cost reduction, and minimizing environmental impact are very much dependent on these technologies involving HVAC systems. HVAC systems can perform better, require less dependency on fossil fuels, and be designed for sustainable building if they accept or generate excess thermal energy and capture and reutilize waste heat.

By heating or cooling a storage medium, TES technology stores thermal energy, which is subsequently used to generate electricity through a heat engine cycle. Its capacity to function in demand-side management is demonstrated by its ability to transfer electrical loads. Both the industrial and residential sectors can make use of technologies that store thermal energy for cooling and heating. Its main parts are a control system, a heat transfer medium, and a thermal storage tank, as shown in Fig. 15.

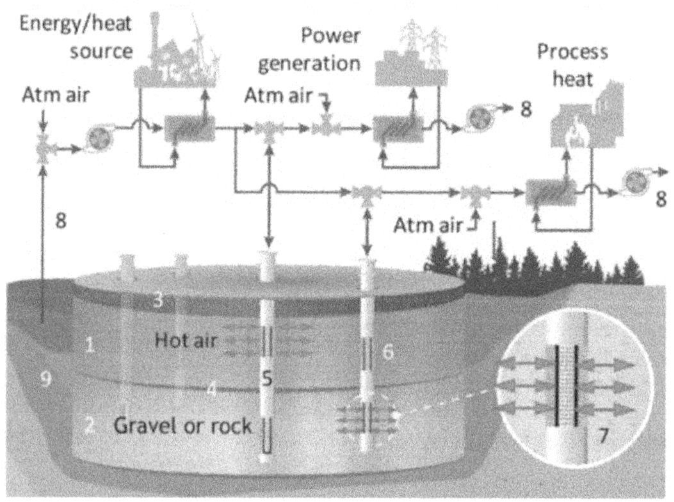

Figure 3.17. Thermal Energy Storage

Source: (Sandia, 2023)

The several kinds of thermal energy storage devices

1. **Sensible Heat Storage:** These systems use different materials, like water, pebbles, or ceramics, to change the temperature of a storage medium to store thermal energy. In these systems, energy and temperature stay proportional, meaning that a material's temperature rises as more energy is applied. Systems like molten

salts, concrete thermal mass systems, and hot water storage tanks frequently use sensible heat TES storage.

2. **Latent Heat Storage:** Latent heat TES devices store thermal energy by basing on the heat the material takes in or releases during the phase change, that is, from a gas to liquid or liquid to solid. The utilization of phase change materials (PCMs) is often used to achieve latent heat storage. These materials can absorb or release large quantities of heat during phase changes, such as solid to liquid or liquid to gas, which is very efficient storage and release of energy. Some of the common PCMs are salt hydrates, paraffin waxes, and some chemical compounds. Latent heat storage is especially common in ice storage systems, where water is frozen as electricity is cheap and thawed to deliver cooling when needed.

3. **Thermochemical Storage:** Thermochemical storage systems use reversible chemical reactions, such as heat absorption or release, to store thermal energy. The energy is stored in materials that can undergo a chemical reaction and is released when the process is reversed. Metal hydrides, metal oxides, and certain salts are examples of common thermochemical materials. In other words, thermochemical systems are systems capable of storing heat through reversible chemical reactions when two substances react or come apart, and absorb or release heat. However, thermochemical storage is the best among the three types of thermal energy Wstorage systems for high-temperature applications because of its large energy capacity in storage and release. They show no energy leakage and a higher energy density (Katlyn Avery, 2024).

Applications of TES in HVAC Systems

1. **Chilled Water Storage:** utilized in large commercial buildings and district cooling systems. It stores chilled water produced during off-peak hours for use during times of high demand. To

save money on power, it moves cooling demands from peak to off-peak times.

2. **Ice Storage Systems:** Cooling during peak hours can be provided by using ice as a storage medium. It is cost-effective for commercial buildings as it reduces peak electricity demand.

3. **Solar Thermal Storage:** Store solar heat for heating and water. It improves the performance of solar-aided HVAC systems.

4. **Industrial Process Heat Storage:** It can store excess thermal energy of industrial processes, reuse them, improve efficiency and reduce operational costs.

3.4.1   Ice and Chilled Water Storage Systems

To effectively control energy demand and manage cooling operations in large buildings and businesses, TES technologies primarily consist of ice and chilled water storage systems. These systems allow storing thermal energy during times of low energy demand (or during off-peak hours) and releasing it to cool in times of higher demand, making them a cost and energy-efficient way of meeting cooling demands as opposed to traditional cooling methods. These systems help lower energy bills, improve system efficiency through peak reduction, shift cooling demand onto nonpeak times, and increase grid stability.

How Ice Storage Systems Work

Ice storage devices allow thermal energy to be stored as ice. Making the ice involves utilizing an electric chiller at off-peak hours, such as nighttime, when electricity costs are cheaper. Once this ice is stored, this ice is then stored in specially designed tanks or bins until needed. The stored ice is used during peak demand periods, to cool air or water for circulating through the building or facility. Instead of the need for direct cooling with electricity, the ice provides cold cooling through the melting of the ice, absorbing heat in the process.

In particular, this technology is useful in large commercial buildings, schools, hospitals, and industrial facilities with high cooling equipment on/off cycles, and high-power costs during peak hours.

- **Chilled Water Storage Systems**

During off-peak hours, the chilled water storage system stores cooling energy as chilled water. is a type of thermal energy storage. As a result, this stored cooling energy is used when demand is at its highest, when taking advantage of its availability, increasing system efficiency, and improving the system's overall operational performance can save energy expenditures. Because these systems draw energy from the earth, they are very useful for buildings and industrial facilities that require high cooling or in instances where lowering energy use during peak usage hours when electricity prices are typically higher is desired.

- **Chilled Water Storage Systems Work**

In cold water storage systems, an auxiliary refrigeration system is chilled with cold water to reach the required temperatures for use in chilling applications later on. The chillers in the systems generate cold water, which is subsequently kept in insulated tanks during off-peak hours. The HVAC system uses the stored chilled water to absorb heat from the room when cooling is required. The chilled water is cooled until it rea During off-peak hours, cold water produced by the systems' chillers is stored in insulated tanks. When cooling is needed, the HVAC system absorbs heat from the space using the chilled water that has been stored. ches a degree where it can be recycled and returned to the storage tank for further chilling.

In large commercial buildings, office complexes and industrial sites, chilled water storage is used frequently. However, it is an excellent solution considering demand-side management in cases of high energy cost or where the peak cooling load needs to be minimized(Rob Ezold, 2024).

What a Chilled Water Storage Systems Work

A chilled water storage system essentially consists of the following essential components.

1. Cooling water and reducing its temperature is done by the chiller. The chiller runs during off-peak such as night when electricity demand is lower to use that chilled water (4–6 degrees C or 39-43 degrees F).

2. Insulated Tank: In an insulated tank, the chilled water is stored before it has to be used. The size of this tank, and thus the moisture contents, will depend on the need for cooling of the facility. It insulates the environment preventing heat gain and keeping chilled water at a low temperature until the time it is used.

Cooling equipment such as "air handling units" (AHUs) and "fan coil units (FCUs) transfer water from a storage tank to the cooling system when the building requires chilled water. The cold water is re-chilled in the tank before it enters the building or process if needed.

Types of Chilled Water Storage Systems

Systems for storing cold water come in two primary varieties:

1. **Ice Storage (Ice-on-Coil or Ice Slurry):** Some systems use ice as a storage medium on the chiller water side of the system, as you may see if you continue reading. Ice that has been frozen around coils or in slurry form melts to provide cooling in the systems under consideration. Hybrid systems may provide superior ice storage and chilled water.

2. **Liquid Water Storage:** It doesn't transform water into ice; instead, it keeps it in an insulated tank. The ice-making process is avoided by pumping cooled water from the ice well into the building's cooling system when needed.

3.4.2   Phase-Change Materials (PCM) in HVAC Applications

Phase change materials, or PCMs, are substances that undergo heating or cooling during melting and solidification at a specific temperature. Heat will be absorbed by a PCM, which will then change from a solid to a liquid and back again. The application of PCMs in thermal energy storage systems and numerous benefits they bring to HVAC temperature control are two of its most appealing qualities. PCMs for HVAC systems in particular are being explored as a sustainable and energy-efficient means of reducing energy consumption, enhancing indoor comfort, and promoting environmentally responsible heating and cooling practices.

Working of PCMs in HVAC Applications

PCMs are integrated as components of the HVAC system itself, of the thermal storage units etc, to maintain temperature and lower energy consumption in all HVAC systems. They are intended to store excess thermal energy at optimal heat periods and release it at lower temperatures to shave off temperature spikes and improve the system performance.

- **PCM:** As the building's temperature rises, PCM absorbs the heat and transforms from a solid to a liquid. Rather, it cools the interior while storing more heat.
- **Heated Release:** The PCM thaws (solidifies) when the outside temperature drops, releasing the heat that has been stored inside to help create a comfortable inside atmosphere without the need for external cooling equipment.

The process of absorbing and then releasing heat maintains indoor temperatures so that heating is not continuously demanded, nor cooling.

Types of Phase-Change Materials for HVAC Systems

Actually, a few kinds of PCMs can be applied to HVAC systems, such as:

- Paraffin and fatty acids are biodegradable, nontoxic organic PCMs. These are used in the applications where they care about the safety and ecological impact. Generally, the latent heat of fusion for organic PCMs is low but they are inexpensive.
- Other PCMs: Salt hydrates are other inorganic materials with a higher latent heat of fusion, which means they can store lower energy for the same volume. Also, they tend to have problems in corrosion and supercooling, thus making them less reliable in some watering systems.
- Mixture of organic and inorganic material: These are the mixtures of organic and inorganic materials to take advantage of both. And they are generally more thermally stable; you can tune them for a certain temperature range that you need for a specific application, for a certain HVAC application(R. K. Sharma et al., 2024).

Benefits of PCMs in HVAC Systems

Additionally, they can lessen the need for mechanical cooling by rerouting energy demand to off-peak hours, when energy is more affordable and accessible. Utilizing stored thermal energy improves HVAC systems by reducing energy use and running expenses.

1. **Improved Indoor Temperature:** PCMs help in maintaining a consistent indoor temperature, which means fewer operations on the air conditioning unit (which can be noisy or disruptive).
2. **Buffered Temperature Fluctuations with Reduced Cooling Load:** Since PCMs can buffer temperature fluctuations, HVAC systems require less cooling during times when temperature fluctuations are high and less cooling loading. By reducing cooling equipment loading and prolonging system lifespan, this is achieved.

3. **PCMs promote** energy saving and carbon saving by lowering the requirement of energy-intensive cooling and heating due to the conventional means that heavily depend on fossil fuels. As an important component of green building technologies, they store and release thermal energy without energy consumption directly.

4. **Energy Usage Reduction and Cost Saving:** Also, PCMs can decrease energy usage and help reduce the wear of the HVAC devices by minimizing the need for continuous operation. Lower maintenance costs and longer system longevity are a result (Kalbasi et al., 2023).

3.4.3 Heat Recovery Strategies in Industrial and Commercial HVAC

The process of recovering waste heat from commercial and industrial systems to save energy is known as heat recovery. Heat recovery techniques are used in commercial and industrial HVAC systems to recover energy that would otherwise be squandered and utilize it to either heat and condition incoming air or power a section of the HVAC system. Heat recovery is a way to reduce businesses' energy consumption, cut down operational costs, and make sustainability easier.

Heat Recovery in HVAC Systems

Heat recovery systems in HVAC setups recover heat from the exhaust air or fluids (e.g. exhaust air from industrial processes, HVAC operation, or refrigeration and transfer the energy into incoming air, water, or other mediums. The type of heat exchanger that can facilitate this exchange will determine the system design and the intended heat exchanger outcome; options include water-to-air, air-to-air, and plate heat exchangers.

The process usually takes place in two stages:

Waste heat (often exhaust air or process fluid) is used to heat up the heat recovery system. There are many sources of this heat, which can be building ventilation systems, machinery, and industrial processes (e.g., refrigeration, ovens, boilers).

Heat Utilization: After being captured, the heat must be moved into a working medium, such as water or fresh air. To eliminate the need for extra energy input from outside sources, the generated energy is then recycled in the HVAC system and used to warm incoming fresh air, preheat water for heating systems, or even support industrial activities.

Types of Heat Recovery Systems in HVAC

1. **HVAC systems in homes and businesses employ "heat recovery ventilation" (HRV)** systems to collect heat from exhaust air and transform it into fresh air. These systems are particularly beneficial for buildings that need continuous ventilation while preserving energy efficiency. A heat exchanger that transfers heat between entering and exiting air streams is used in the HRV system to control the inside temperature and reduce the energy required for heating or cooling.

2. **Energy Recovery Ventilation (ERV):** An ERV system is an HRV system that recovers both heat and humidity. An ERV system can lighten the task of dehumidification systems by transferring moisture as well as heat, allowing temperature and humidity control to be done more efficiently. With this it makes ERVs ideal for places where there are temperature and moisture issues combined.

3. **Heat Recovery Chillers:** They are made for recovering industrial processes or HVAC unit waste heat to usable. A wide range of industrial equipment can be heated using the recovered heat, as well as water and spaces thermal power. Depending on the use, the recovered heat might be used to heat water, rooms, or other industrial machinery. Larger industrial or commercial applications where high-capacity cooling, as well as waste heat, are available for recovery will often use heat recovery chillers.

Heat Recovery from Refrigeration Systems: In cold storage, food processing, refrigeration and pharmaceutical facilities, industries that depend on refrigeration systems for cooling applications, the refrigeration systems generate waste heat, which can be recovered and utilized. To use some of the surplus heat from refrigeration compressors for space heating or hot water generation, a heat recovery system can be installed.

Challenges of Implementing Heat Recovery in HVAC

- **High Upfront Costs:** A heat recovery system's initial installation costs are considerable and particularly important for sophisticated industrial applications or existing buildings that need to be retrofitted. Such a cost can put off some businesses from embracing these technologies even though the long-term savings are significant.

- **Space Requirement:** Since heat recovery systems, especially large-scale systems such as heat exchangers or heat recovery chillers, require so much space to be built, it might be limiting. This may also be a challenge for incorporating heat recovery solutions in facilities constrained by space.

- Additionally, a heat recovery system has to be added to a configured HVAC system meaning it is complex or complex design and installation. Careful engineering and integration are needed to allow the recovered heat to be used effectively so as to not overload the system and minimize the impact on overall performance.

- **Concluding Remarks:** The heat recovery systems reduce energy use but maintaining them will keep them functioning well. Heat exchangers, fans, and pumps are all components that need to be cleaned and serviced from time to time to avoid degradation of performance over the course of time.

3.5 Chapter Synthesis

Heat recovery techniques may be integrated in the manufacturing and institutional sectors based on numerous operating characteristics and

HVAC systems can take advantage of phase change materials and TES to improve their energy efficiency as well as their sustainability. By allowing the phase change materials or the thermal energy storage in HVAC systems to be thermally energized during off-peak hours, the off-peak usage of TES storage improves the transport energy use and energy efficiency of the system. The energy loads are stored with TES during off-peak periods to make them more balanced and thus reduce costs. Aside from this, the use of PCMs further improves stability of temperature control by storing and releasing heat during phase transitions, reducing the need for traditional heating and cooling systems. It guarantees a comfortable home atmosphere and improves energy efficiency and operating costs. Heat recovery systems, including "Heat Recovery Ventilation" (HRV) and Energy Recovery Ventilation (ERV), provide even more efficient use of heat by reusing it to heat water or other locations. Even while these systems offer significant long-term savings, they do need a challenging initial investment and modification of current systems. Economic and environmental benefits, however, exceed these challenges, especially when taking into account new energy and climate supply methods that seek to reduce dependency on fossil fuels and GHG emissions. When combined, the above technologies help create a more sustainable future through efficient usage of energy, reduced operational costs, and improved overall performance of HVAC systems. These innovations continue to develop and projected to be adopted for making buildings and the industrial sector even more energy-efficient and less environmentally impactful.

Multiple Choice Questions (MCQs)

1. **What is a primary advantage of solar-assisted HVAC systems?**

    a. They operate only during daylight hours
    b. They reduce dependency on fossil fuels
    c. They increase HVAC operational costs
    d. They require no maintenance

2. **Which factor is most crucial in designing an efficient geothermal heat pump system?**

    a. The length of the refrigerant pipes
    b. The efficiency of the evaporator coil
    c. The thermal conductivity of the ground
    d. The humidity levels in the building

3. **Which of the following best describes a hybrid HVAC system?**

    a. A system that combines two or more energy sources for heating and cooling
    b. A system that operates on battery backup only
    c. A system that requires constant human intervention
    d. A system that uses only renewable energy sources

4. **Why is the transition from hydrofluorocarbons (HFCs) to low-GWP refrigerants important?**

    a. HFCs are more energy-efficient than alternatives
    b. HFCs have a high global warming potential
    c. Low-GWP refrigerants are more expensive
    d. HFCs improve cooling performance in HVAC systems

5. **Which of the following is an example of an emerging eco-friendly cooling technology?**

    a. Absorption cooling using waste heat
    b. Window air conditioners with R-410A refrigerant
    c. Traditional vapor-compression cooling
    d. High-energy consumption cooling towers

6. **What is a key function of regulatory frameworks for sustainable refrigerants?**

    a. To phase out inefficient cooling systems
    b. To ban all air conditioning units
    c. To encourage the use of high-GWP refrigerants
    d. To increase HVAC operational costs

7. **How do ice and chilled water storage systems contribute to HVAC sustainability?**

    a. By storing excess thermal energy for later use
    b. By reducing HVAC efficiency
    c. By increasing peak electricity demand
    d. By eliminating the need for cooling in buildings

8. **Which material is commonly used in phase-change materials (PCMs) for HVAC applications?**

    a. Concrete
    b. Paraffin wax
    c. Copper
    d. Fiberglass

9. **What is a primary benefit of heat recovery strategies in industrial HVAC systems?**

    a. They increase overall energy efficiency by reusing waste heat
    b. They require frequent repairs and replacements
    c. They consume more energy than traditional HVAC systems
    d. They work only in residential applications

10. **Which approach best represents a sustainable HVAC strategy?**

    a. Using only traditional refrigerants
    b. Increasing energy consumption for better performance
    c. Integrating renewable energy sources and energy-efficient technologies
    d. Ignoring regulatory frameworks on refrigerants

Answers

1	2	3	4	5	6	7	8	9	10
b	c	a	b	a	a	a	b	a	c

CHAPTER 04

ENERGY AUDITS, PERFORMANCE OPTIMIZATION, AND CASE STUDIES

4.1   Chapter Overview

In this chapter, we explore actionability of energy audits to analyze energy consumption patterns, identify inefficiencies, and define the ways that can lead to the improved performance in industrial or commercial sector. Energy audit is a critical diagnosis tool that shows the amount of energy used across operations, the comparison with industry standards and identification of a potential cost saving. Audits in turn pattern the processing of energy inputs and outputs so that organizations can improve energy efficiency, reduce operating expenses and keep up with sustainability goals. The chapter classifies the audits as the preliminary, general and detailed and explains that there are differences in scope and depth of audits and methodologies of data collection, analysis and reporting (P. Sharma et al., 2021).

After the audit process, the chapter focuses on strategies for optimizing performance based on the audit findings to energy efficiency. The strategies being used for that include equipment retrofitting, process improvements, demand-side management, and renewable energy integration. It also talks about technologies that allow real time monitoring and prediction of maintenance such as energy management systems (EMS), smart metering and automation. Finally, the main purpose of the chapter is to draw attention toward the regulatory compliance, the financial incentives, and policy framework for making businesses to use energy efficiency practices. In addition, it elaborates on the implementation impediments, such as the costs, technological restrictions, and reluctance to alter things (Kumar Singh et al., 2023).

This case demonstrates the successful implementation of the energy audit and performance optimization initiatives on the basis of different industries as described in the chapter. Thus these cases illustrate how the companies have identified opportunities for inefficiency, adapted the corrective measures and tangible improvement of energy consumption and cost reduction. Those who set up energy efficiency programs also provide practical lessons about what pays off best practice, return on investment (ROI) considerations and long-term sustainability benefits. This chapter finally calls for continuous improvement approach towards conducting energy audits that will treat energy audits as a continuous process with an aim of having sustained in efficiency gains and congruent with the energy regulations and market trends (ECBC, 2001).

Energy auditing is a must for Indian industries to assess their energy consumption, find out the areas of inefficiency and rectify them by taking up energy saving measures. With rising costs of energy, need for sustainability becoming increasingly important, Energy Audits have become needed for any business to cut down on energy usage, reduce operating expenses, and curb environmental impact. Energy audits are very important in India where industries are major contributors of energy consumption and greenhouse gas emissions and contribute to the sustainability of business and global competitiveness.

Energy audits in Indian industries are a detailed study of the energy consumed in the industries in terms of electricity, fuel, etc. In the normal case, the audit process entails a detailed analysis of energy bills; on site visits to locate energy waste; and the formulation of suggestions for energy efficiency. The uses of specialized tools and employed techniques of energy auditors help them to measure energy consumption, detect inefficiencies present in equipment and process, and propose for least cost means of reducing energy usage.

Indian industries benefit from conducting energy audits of its energy consumption patterns and take proactive measures to optimize their energy use (Elion, 2024).

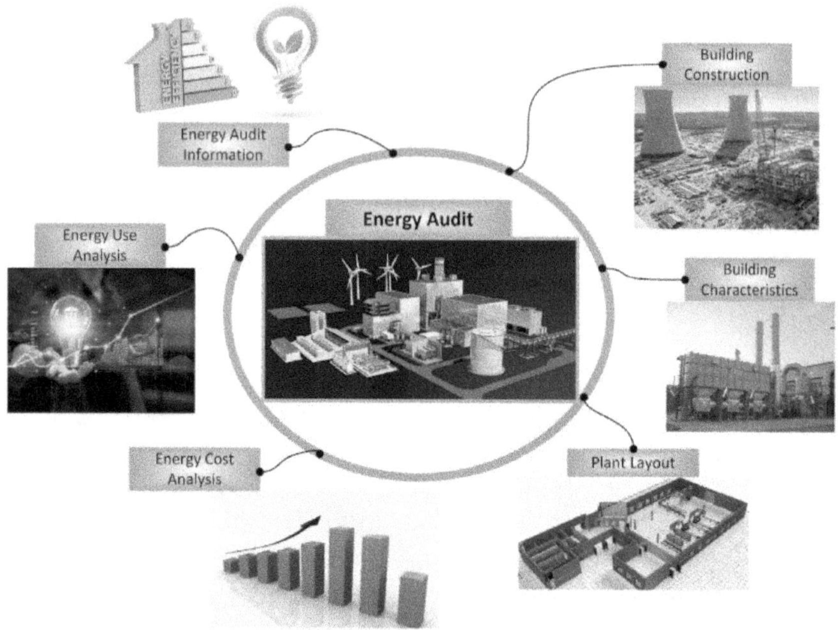

Figure 4.1: Energy Audits

Source: (Al Momani et al., 2023)

The figure 4.1 shows the multifaceted process in an energy audit for energy efficiency in building construction. This document the relatedness of energy use, cost, plant layout to fundamental building characteristics and construction methods. The lightbulb and the tiered energy efficiency graphic reflect the finding of the audit to create strategy for reducing energy consumption and costs.

Benefits of Energy Audits for Indian

- **Financial Advantages:** Businesses can reduce their cost by identifying the areas that waste energy and take measures to make efforts for better energy efficiency. Thus, Indian industries become more competitive in the global market owing to the improved operational efficiency and lower operating costs.

- **Environmental Benefits:** Energy audits assist in reduction of energy footprint on the part of business by reducing greenhouse gas emissions as well as saving natural resources. Not only this adds to the future more sustainable but this also makes the companies comply with relevant regulations and standards.

- **Long-Term Sustainability and Compliance:** Energy audits enable businesses to make strategic investments in sustainable technologies that yield long-term benefits for their operations. By addressing energy efficiency requirements through audits, businesses can ensure compliance with relevant regulations and standards, ultimately leading to a more sustainable and environmentally responsible industry (Elion, 2024).

4.2   Conducting HVAC Energy Audits

HVAC energy audit is a measure of efficiency of heating, ventilation, and air conditioning systems. It is this process that finds how much energy these systems use and where the improvement can be made. An HVAC energy audit should be conducted in order to keep the system efficient. This allows you to detect issues that will lead to wasted energy and increased costs.

This process requires a critical role to be played by an HVAC energy auditor. It analyses the performance of the HVAC system at present. Components that the auditor looks into include ductwork, insulation and equipment settings. This allows them to find possible leakages or inefficiencies in a home energy audit and an energy efficiency audit.

The auditor offers useful information on how to improve system performance. Varied types often have specialized tools for measuring airflow and temperature differences during a home energy audit or energy efficiency audit. Thus these findings allow the homeowners to understand their energy consumption (Service, 2024).

Importance of Regular Audits

Regular audits contribute significantly to building sustainability. They help in identifying areas for improvement that can reduce energy usage. In such an environment, where the energy costs are rising, it's imperative to become energy efficient as homeowners and businesses.

Homeowners benefit from routine home energy audits by saving money over time. A comprehensive residential energy audit can reveal simple fixes that improve efficiency without major investments. For example, a home energy audit shows that sealing duct leaks can enhance airflow and reduce heating or cooling needs during an energy efficiency audit.

Furthermore, consistent auditing supports broader energy management goals. Many organizations implement an energy audit program to track performance over time. This ongoing assessment fosters accountability and encourages ongoing improvements.

Energy Audit Findings

The findings from an HVAC energy audit guide decision-making. Homeowners receive reports detailing current system performance and recommendations for upgrades. These may include a home energy audit, replacing outdated equipment with more efficient models or improving insulation.

The results can highlight the need for regular maintenance checks. An effective energy auditing process leads to fewer breakdowns and longer equipment lifespan. This proactive approach ensures that systems operate efficiently throughout their service life.

Benefits of Auditing

Conducting an HVAC audit has multiple benefits. It can extend the life of the system by identifying maintenance needs early. Regular audits can also enhance energy efficiency, leading to lower utility bills.

Furthermore, audits promote better indoor air quality, which is essential for health. Improved comfort levels contribute to a more pleasant living or working environment (Service, 2024).

Types of Energy Audits

- **Basic Audit:** A **basic energy audit** is often the first step for many building owners. This type involves a simple walkthrough of the facility. Auditors look for obvious energy waste and inefficiencies. They assess lighting, insulation, and HVAC systems. Data collected includes an overview of current energy use. This audit helps identify quick fixes that can reduce costs. For small buildings or homes, a basic audit is usually sufficient. Owners benefit from immediate insights without extensive investment.

- **Detailed Audit:** A **detailed energy audit** goes deeper into energy use. It requires more comprehensive data collection. Auditors analyze equipment performance and operational practices. They may use specialized tools to measure energy consumption accurately. This level is suitable for larger facilities or those with high energy bills. If detailed audit is done, then the picture becomes clearer of where there are savings to be made. Owners of the building receive a report highlighting the recommendations, as well as possible savings.

- **Comprehensive Audit:** This is the most comprehensive of the options. It analyses all aspects of energy usage of a building. It is evaluating mechanical systems, the building envelope, the occupant behaviour. Over time, auditors use the advanced monitoring software, and collect extensive data. In particular, they look at how various systems influence the overall efficiency. Large industrial facilities or organizations that are serious about cutting energy use would find this audit most appropriate.

- **Industrial Audit:** Manufacturing and production facilities are the scope of industrial energy audits. These audits deal with unique challenges that are faced by the industries. Machines and processes are evaluated as well as operational practices. Production schedules, and machine efficiency rating data are collected. The aim is to find areas of possible waste reduction and increased productivity. These targeted assessments are of great benefit to owners of industrial sites.

- **Free Energy Audits:** As part of their offerings, numerous utility companies provide free energy audits. These programs encourage customers to improve their energy efficiency. While these audits may not be as thorough, they provide valuable insights. Owners can learn about available rebates or incentives for upgrades. Free audits are particularly beneficial for residential properties seeking immediate solutions.

- **Quality Audit Firms:** Choosing a reputable energy audit firm is crucial for accurate results. A good firm uses proven methodologies and experienced auditors. They should offer various services, including site assessments and reporting. Quality firms also utilize advanced energy auditing software platforms to enhance their analysis. This technology aids in providing detailed reports tailored to specific building needs (Service, 2024).

4.3   Performance Monitoring and Predictive Maintenance

The performance monitoring and the predictive maintenance are important strategies for ensuring the optimal operation of energy systems in industrial and commercial setting. In contrast, these approaches rely on real time data collection with the use of advanced analytics and machine learning subject matters to detect the inefficiencies, the failures, and the performance in general. The International Energy Agency along with others estimate that predictive maintenance can cut down the unplanned downtime by as much as 30 to 50 and equipment life by 20 to 40%. The advent of the industry 4.0 with internet of things (IoT) has opened doors for the businesses to integrate smart sensors, cloud computing, and the AI powered analytics of the energy efficiency and minimizing operational disruptions (Ryalat et al., 2024).

Performance Monitoring

Most people agree that one of the most crucial elements of a continuous performance management strategy is performance monitoring. In order to improve employee performance, it is the process of making accurate and measurable observations based on a comparison of actual and desired performance results as stated in the performance appraisal. It allows for the completion of the training and education needs that were emphasized in the performance review. Performance monitoring makes it possible to assign better performance delivery methods and procedures. Additionally, it allows the performance criteria to be redefined for use as performance standards. Lastly, it enables the system to adapt to required modifications brought about by shifting business circumstances. Employee commitment in terms of following agreed-upon performance improvement measures is required as part of the monitoring process, which also include identifying gaps between planned and actual performance and thoroughly discussing and developing an improvement plan.

Process of Performance Monitoring:

There is a methodical approach that must be followed in order to establish performance monitoring within the company. The steps in performance monitoring are as follows:

Gap identification: A manager must regularly review the performance plan with the staff during the appraisal period. Making employees aware of the required level of performance that must be provided is the goal of the performance plan discussion. Employees can also anticipate how and on what basis their performance will be evaluated. Actual performance is compared to the predetermined level of performance through the use of an efficient performance appraisal system. Additionally, it indicates the target performance level for the upcoming assessment cycle as well as the strengths, areas for development, and places for improvement. As there exists many factors to be monitored, a manager has to take decision on what needs to be monitored. After that a procedure is defined about the way of obtaining the data regarding the identified monitoring factors. With the help of performance appraisal vital information related to various aspects of performance could be gathered. This data serves as a source for identifying performance shortcomings. It is critical that a manager at this point takes into account every element that could affect an employee's performance. The direction and extent of the influence on the performance must also be taken into account.

Discussion, feedback, planning: The next step for a manager is to create a system for informing the relevant staff members about their areas of strength and growth after they have a good understanding of them. Employee participation in the performance monitoring activity is the goal of exchanging performance-related information. Employee performance is tracked, evaluated, and audited with their cooperation. The employee and the performance review are only discussed after this phase. It's critical to recognize that performance reviews are more than just summarizing what has been or ought to have been done. It has to do with letting an employee know how he is doing and what effects that performance will have on him in the future. Additionally, the effects of an employee's performance are examined in relation to the team and organization (Pathshala, 2024).

Performance Monitoring: Key Components

Performance monitoring is a constant test of energy usage, system efficiency and operational parameters to prove that equipment is working well. This process typically includes:

- **Real-time data acquisition** – Collecting the performance metrics by means of smart meters, IoT sensors and Building Management Systems (BMS).
- **Energy benchmarking** – Determining how the UNR business is compared to industry standards and historical data to see where possible energy consumption efficiencies can be gained.

- **Fault detection and diagnostics (FDD)** – A way to use automated systems to detect performance anomaly and deviation.
- **Load analysis and optimization** – To realize maximum energy efficiency, such as demand patterns tuned to adjust operations (Jasim et al., 2023).

Predictive maintenance

Predictive maintenance is an approach to maintenance that puts the emphasis on hearing what your enterprise assets are trying to tell you. The machines in your factories, your fleet of trucks, your industrial equipment – they've been talking to you for years. If you can listen closely, you can understand when your machines are about to break down and what they need to run longer and more smoothly.)

Your company may plan maintenance for when and where it's most urgently needed by using predictive maintenance to foresee equipment problems. It gives you the knowledge you need to operate your assets at their best without overtaxing them and running the risk of expensive malfunctions. Predictive maintenance reduces equipment failure and downtime and increases the lifespan of critical assets by connecting IoT-enabled enterprise assets, applying advanced analytics to the real-time data they generate, and using the insights gathered to inform economical, effective maintenance protocols (SAP, 2024).

Predictive Maintenance: Role and Benefits

Predictive maintenance or sometimes represented as PdM is an application of use of AI, machine learning, big data analytics to foresee machine failures beforehand. PdM is based on real time operation data and is therefore condition-based and not a traditional preventive maintenance type that is schedule based.

Key advantages include:

- **Reduced downtime and maintenance costs** – They also say that predictive maintenance can cut maintenance costs by up to 20–30 percent and avoid surprises failures.
- **Extended equipment lifespan** – This is viable because continuous monitoring will provide the early detection of wear and degradation of components.
- **Energy efficiency improvements** – 5–15% energy savings are possible to realize by identifying and resolving performance problems in HVAC systems, motors and industrial equipment.
- **Workforce optimization** – Those in maintenance teams can shift to a proactive intervention, thereby improving productivity.

4.3.1 Role of Data Analytics in HVAC Energy Management

An HVAC energy management system is an integrated solution to control and optimize the performance of heating, ventilation, and air conditioning systems within a building. Advanced technologies, such as sensors, controllers and software, are used in these systems to monitor and change HVAC operations on real real-time basis. That way, they make sure that the building's environment isn't too hot or too cold, uses as little energy as possible, and saves time and money.

HVAC energy management systems collect data from a number of sensors placed throughout the building. The sensors monitor the key parameters like temperature, humidity, air quality and occupancy rates. The system's software processes the data and then performs algorithms to analyze the information and adjust the HVAC equipment. For instance, if the sensors indicate that a room is unoccupied, the system can decrease heating or cooling for that area in order to lower the utility costs. On the other hand, if occupancy is detected, the system can set the HVAC settings to conform to comfort (Aemaco, 2024).

HVAC (Heating, Ventilation, and Air Conditioning) system is a must component for modern buildings that ensure air and temperature control. However, these systems consume a great amount of energy, approximately 40% of the total amount of energy used in a building. It is possible to manage Building energy in an effective way through HVAC energy management, as it can reduce building energy cost and improve building performance. (SBA, 2024).

Here are five strategies for optimizing HVAC systems:

1. **Regular Maintenance:** Having regular maintenance is essential for the HVAC system to be running at its best. Periodic cleaning, checking, and replacement of filters and cleaning and tuning up of the system can prevent inefficiencies and breakdowns.

2. **Upgrading Equipment:** in upgrading to more energy efficient HVAC equipment, energy consumption and costs can be significantly reduced. Find equipment that is ENERGY STAR certified and opt for systems that have high Seasonal Energy Efficiency Ratios (SEER) and Energy Efficiency Ratios (EER).

3. **Implementing Building Automation Systems:** Automation systems of building (BAS), control and monitor HVAC climate control in the entire complex in one place, helping you save energy by automatically switching to lower temperature and ventilation mode when possible. For instance, use of occupancy sensors would eliminate energy consumption by automatically turning on heating and cooling when a space is not occupied.

4. **Sealing Ducts:** Exactly 30% of a building's heating and cooling energy losses can be accounted for by poorly sealed ducts. Sealing ducts can increase the efficiency of HVAC systems and thereby reduce energy consumption and costs.

5. **Optimizing Settings:** Reducing energy consumption can also be achieved with optimizing temperature settings. This can result in varying degrees higher or lower in temperature during the

summer and in the winter and have a notable effect on energy consumption and costs (SBA, 2024).

Figure 4.2: HVAC Energy Management Systems

Source: (Indiamart, 2024)

Benefits of HVAC Energy Management Systems

- **Energy Efficiency:** The one of the major benefits of HVAC energy management systems is that they help to improve energy efficiency. These systems monitor and optimize HVAC operations continuously to reduce the energy consumption by a big margin.

This also enables to decrease the utility bills and for environmental sustainability by lowering the building's carbon footprint.

- **Cost Savings:** Cost savings resulting from HVAC energy management system energy savings are directly related. These systems will result in substantial monthly reductions in the business energy bills, and it is a worthwhile investment. In addition, utility companies provide incentives and rebates for the building that implements energy efficient technologies, and generates additional return on investment.

- **Improved Comfort and Productivity:** A good managed HVAC system will provide consistently comfortable indoor environment for occupants of a building, which is essential for their well-being and their productivity. They have found that employees tend to work better and suffer less from health problems in climates with stable and optimal temperature and air quality.

- **Enhanced Control and Flexibility:** With modern HVAC energy management systems building managers have more control and flexibility of their HVAC operations. Even these systems typically have user interfaces and mobile applications which are user friendly and for remote monitoring and switching over. It implies that managers can change on the go such that HVAC system functions efficiently at all times (Aemaco, 2024).

4.3.2 Fault Detection and Diagnostics in HVAC Equipment

The Heating, Ventilation and Air Conditioning (HVAC) systems are the foundations to the inside environment, the room temperature, the air quality and the energy consumption. But most unfortunately, faults in HVAC equipment can cause increased energy consumption, system ineffectiveness, poor thermal performance and high maintenance costs. Fault Detection and Diagnostics (FDD) is an approach to identify, analyze and repair operational anomaly in HVAC systems. FDD uses high technologies such as advanced sensors, machine learning algorithms, and on real time monitoring to improve the reliability and performance of HVAC systems and reduce downtime and reduction of energy waste (V. Sharma & Mistry, 2023).

With extensive usage of digital technologies within maintenance and operation processes for HVAC systems, the integration into systems has become necessary. Reactive and preventive maintenance strategies are not very effective in exposing hidden inefficiencies or in predicting whether a failure is going to occur before it manifests. Changes in smart building technology, Internet of Things linked sensors, and cloud-based analytics have immensely increased the ability of a facility manager to pinpoint and formulate the fault in actual time. Besides increasing energy efficiency, having an efficient FDD system also extends the equipment's lifespan, guarantee compliance to environmental regulations and reduce unexpected breakdowns.

The modern HVAC fault detection systems make use of automated diagnostic algorithms, data driven insights, as well as the analytics from AI to continuously monitor the system. Such systems look into large volumes of data, identify irregular pattern and create actionable suggestions to enhance performance. In the same vein, the integration of FDD with predictive maintenance strategies facilitates proactive interventions in order to minimize overall operational costs and operationalize the strategies of sustainability efforts.

In this chapter, a few faults that may occur in HVAC and their detection methods, diagnostic techniques, implementation strategies and real-life case study are explored. Knowledge of the meaning of FDD in HVAC systems allows stakeholders to create more efficient maintenance strategies that improving energy efficiency, operational risks and ensuring long term system reliability (Hosseini Gourabpasi & Nik-Bakht, 2024).

Types of HVAC Faults

Types of faults in an HVAC system can be generally categorized:

- **System Response Errors: It** refers to sensing faults such as inaccurate readings, sensor drift, and sensor failure thus creating incorrect system responses. Environmental factors, calibration errors or age of the equipment may lead to such problems.
- **System Efficiency:** Faults in the components like the compressor malfunctioning, fan motor failing, clogged air filters, and the leaking of the refrigerant effect system efficiency. However, most of these faults result in overheating, inefficiency in cooling, as well as wear and tear of the system component.

While some of these common controls system faults define how the system operates, others result in more energy use than necessary. Such issues may also be caused by automation errors and software bugs.

Fault Detection Methods

There are many techniques used to find faults in HVAC equipment.

1. Rule-Based Fault Detection

- This method defines predefined rules and thresholds to evaluate system operation to detect abnormal behaviour. Examples include:
- The threshold is set to trigger an alert when temperature readings exceed above or falls below.
- A fan failure may also be indicated if airflow velocity declines below acceptable limits.
- Cycling of the compressor beyond what would be normal can means refrigerant leaks or thermostat failure.

2. Model-Based Fault Detection

This method relies on mathematical or physics-based model for comparing the expected and actual performance. Predicted values are assumed and for any deviation from them, a potential fault is indicated. These models consider:

- Heat transfer and thermodynamic principles
- System-specific performance benchmarks
- Deviation analysis for predictive maintenance

3. Data-Driven Fault Detection

This approach exploits the machine learning algorithms and historical data in order to predict patterns and anomalies of HVAC performance. It includes:

- **Supervised Learning:** The labelled fault data can be used for training models which can predict specific failures.
- **Unsupervised Learning:** Clustering and anomaly detection clustering is the method which detects unusual patterns without predefined labels.
- **Deep Learning Techniques:** Neural network is applied to detect the complex patterns in large data sets.

4.3.3   Preventive vs. Predictive Maintenance: Which is Better?

Preventive maintenance programs plan routine maintenance chores at regular periods in advance and use past data to predict the expected condition of an asset. Although this helps with planning, since most asset failures are unanticipated, assets may be over- or under-maintained. For instance, a problem may be identified too late to stop harm to an asset, which will probably result in more downtime while it is addressed, or time and money may be wasted.

Predictive maintenance recognizes the true state of the equipment and prevents needless repair. This implies that it can identify and address issues before more significant ones arise, even before preventive maintenance.

In order to produce insights, predictive maintenance makes use of emerging technologies such as artificial intelligence, machine learning, and the Internet of Things (IoT). Corrective maintenance work orders are automatically generated by maintenance management systems and software, allowing maintenance teams, data scientists, and other staff members to make quicker, better informed, and more cost-effective decisions.

Inventory management processes, such as manpower and replacement parts supply chains, become more sustainable and efficient by reducing waste and energy consumption. Real-time analytics, such as digital twins, which can be used to simulate scenarios and other maintenance alternatives without endangering production, can be used to input data into predictive maintenance and other maintenance practices.

There are obstacles to overcome for predictive maintenance to be effective or even possible, such as complexity, training and data. Predictive maintenance requires a modern data and systems infrastructure that may make it costly to set up when compared with preventive maintenance. It can be costly and time-consuming to train employees to use the new procedures and tools and to properly evaluate data. Additionally, predictive maintenance depends on gathering large amounts of particular data. Finally, a cultural shift is necessary to support the transition from planned to more flexible daily operations, which can be difficult when implementing a predictive maintenance strategy.

In conclusion, predictive and preventive maintenance approaches are quite different even though they both aim to improve asset dependability and lower failure risk. While predictive maintenance is on giving the appropriate information about particular assets at the appropriate time, preventive maintenance is consistent and routine. While predictive maintenance might be more beneficial for strategic assets where failure is less predictable and has a greater business impact, preventive maintenance is best suited for assets with predictable failure patterns (such as reoccurring or frequent issues) and relatively low impact from failure. In the end, effective implementation and operation of predictive maintenance techniques will lead to satisfied clients and significant cost savings through improved asset performance and maintenance (Gulvin, 2023).

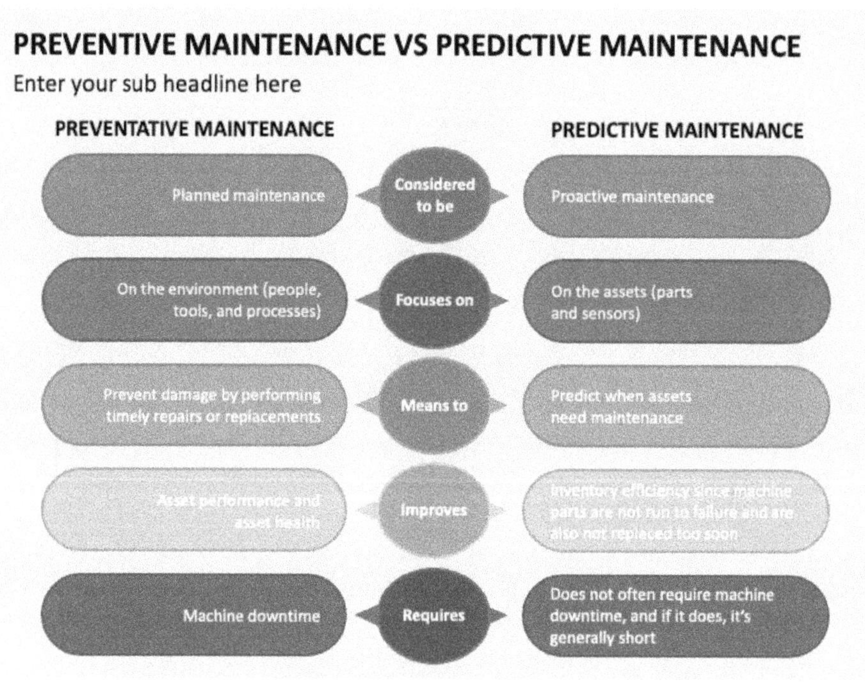

Figure 4.3: Preventive vs. Predictive Maintenance

Source: (Sketchbubble, 2024)

Figure 4.3 visually compares preventative and predictive maintenance, highlighting that preventative maintenance is planned and focuses on the environment to prevent damage through timely repairs, impacting asset performance and requiring machine downtime. In contrast, predictive maintenance is proactive, focuses on assets to predict maintenance needs, improves inventory efficiency by avoiding premature replacements, and minimizes machine downtime.

Preventive Maintenance

Preventive maintenance involves performing normal maintenance work at regular intervals to lower the likelihood that an asset may break down. Downtime is scheduled using historical averages, such as mean-time-between-failures (MTBF), and best practices. Although they have existed since around 1900, preventive maintenance techniques have gained popularity since the late 1950s (link sits outside ibm.com).

There are now three basic forms of preventative maintenance that all entail performing routine maintenance, but they range in scheduling and are designed to serve various business objectives.

- **Usage-based preventive maintenance** regimens, such as replacing your car's tires after 50,000 miles, base upcoming inspections and maintenance on asset usage.
- **calendar-based preventative maintenance** establishes precise maintenance schedules, such yearly furnace service for your house.
- **Condition-based maintenance** develops schedules according to elements such as asset deterioration and wear.

All forms of preventive maintenance involve scheduling machine downtime in advance and using checklists for maintenance tasks such as inspections, repairs, cleanings, adjustments, and replacements.

Predictive Maintenance

Predictive maintenance continuously evaluates the state of an asset, building on condition-based monitoring. AI-enabled enterprise asset management (EAM), computerised maintenance management systems (CMMS), and other maintenance software get real-time data collection from sensors. Advanced data analysis tools and procedures like machine learning (ML) may recognise, detect, and resolve problems as they arise with the use of these kinds of software. The risk of the asset degrading later on is reduced by using algorithms to create models that forecast when possible issues might occur in the future. As a result, maintenance expenses may decrease, downtime may be reduced by approximately 35–50%, and lifespan may rise by 20–40% (link lives outside ibm.com).

A variety of condition monitoring methods, such as vibration analysis, motor circuit analysis, temperature (thermal), lubrication (oil, fluids), sound (ultrasonic acoustics), and others, are employed to detect anomalies in assets and give early indications of possible issues. For instance, an increase in a component's temperature may be a sign of coolant or airflow obstruction; strange vibrations may be a sign of wear and tear or misaligned moving parts; and variations in sound might serve as early indicators of flaws that are imperceptible to the human ear.

Predictive maintenance was first used by the oil and gas sector to reduce the risk of environmental catastrophes, and other sectors are starting to realise its advantages. Undiscovered food storage problems, for instance, could have serious health repercussions in the food and beverage sector. In the shipping business, foreseeing and preventing equipment failures lowers the amount of repairs that must be conducted at sea, when they are more difficult and costly than in port (Gulvin, 2023).

Organizations without advanced data systems and less complex equipment can execute preventive maintenance as an affordable and applicable solution.

Predictive maintenance works best for systems featuring critical high-value equipment since it helps minimize time without operation while delivering peak system performance.

The combination of improved efficiency alongside reduction of expenses has made predictive maintenance the preferred solution for manufacturing and utilities and aviation organizations. Most organizations achieve optimal results by combining preventive maintenance with predictive maintenance strategies between their low-risk assets and crucial systems (Click maint, 2023).

4.4 Retrofitting and Upgrading HVAC Systems for Energy Efficiency

In many types of buildings, the Heating, Ventilation, and Air Conditioning (HVAC) system is essential to achieving thermal comfort and indoor air quality. Nevertheless, most older HVAC systems have inefficiencies that facilitate excessive energy usage, increase operation cost, and boost the environmental burden. With the growing awareness of the importance of switching to more energy-efficient equipment's and magazines such as the Green Building Journal, upgrading and retrofitting of HVAC systems have emerged as indispensable ways of boosting energy efficiency, decreasing the carbon footprints, and adhering to the evolving energy standards.

In light of global energy demand, building owners and facility managers are looking for ways to improve HVAC system performance and minimize operating costs. According to the International Energy Agency (IEA), buildings utilize about 40% of the energy produced worldwide, with HVAC systems accounting for 20% of that total. Retrofitting becomes an essential part of sustainability efforts as slower HVAC systems do not only increase electricity costs but also increase greenhouse gas emissions.

Retrofitting of HVAC systems means modernizing the old equipment, smart technologies and optimization of energy consumption to make them feasible with modern standards. Simply sealing duct leaks, upgrading thermostats or even replacing outdated chillers, can all be included in this process. In addition, artificial intelligence has also evolved with the advancements in the Internet of Things and predictive maintenance as they all now revolutionized how the HVAC systems work and that includes real-time monitoring, adaptive controls and automated fault detection.

Improving the energy efficiency and cost savings through upgrading of HVAC systems also improves the indoor air quality and the occupant comfort and system longevity. Such transition is also provoked by governmental regulations and financial incentives for improving energy efficiency of HVAC, which finally bring us close to high performance, environmentally friendly HVAC solutions. HVAC retrofits provide an increased property value, reduced carbon footprint, as well as an 'Energy Star' or 'LEED' energy efficiency compliance.

This chapter describes the different aspects of the CHP retrofitting's such as benefits, key components to retrofit, technologies available, strategies for implementation, and real case studies. An understanding of the need for retrofitting allows the stakeholders to plan for the optimal performance of HVAC through sustainability initiatives (TERI, 2019).

The Need for Retrofitting HVAC Systems

- **Energy Consumption Reduction:** Modern energy efficient HVAC systems consume only 30 to 50 percent more energy than older systems. Retrofitting diminishes the power demand, as well as the operational costs.
- **Regulatory Compliance:** Today, governments worldwide have begun to upgrade HVAC systems, and this requires both upgrades and stricter energy efficiency regulations (e.g., ASHRAE standards, ISO 50001, Energy Star requirements) where they exist.
- **Environmental Impact:** Transiting to eco-friendly refrigerants and energy efficient systems will lead to reduction of greenhouse gas emissions.
- **Enhanced Indoor Air Quality (IAQ):** improving the air quality first to guarantee both occupant health and good air quality, especially in places that are used sporadically (such as offices, factories, and warehouses during the day) or close to chemical activities that produce ammonia.
- **Extended Equipment Lifespan:** Retrofitting can slow down the replacement of equipment that would otherwise need to be replaced more frequently resulting in a longer useful operational life for the system.

Key Components for Retrofitting and Upgrading

- **High-Efficiency Compressors:** Modern variable speed compressors change the cooling output based on changing demand – reducing waste energy by orders of magnitude.
- **Smart Thermostats and Controls:** The advanced thermostats with AI based algorithms adjust the settings and reduce the energy that is not required.
- **Variable Frequency Drives (VFDs):** HVAC components are made up to operate at variable speeds by installing VFDs on motors and fans.
- **Energy Recovery Ventilation (ERV) Systems:** To precondition the incoming air, exhaust air energy is captured and reused.
- **Upgraded Insulation and Duct Sealing:** This cut down on-air leaks throughout the ducts and improves the system's efficiency and performance.
- **Use of Eco-Friendly Refrigerants:** Changing from the use of older refrigerants (e.g. R-22) to green ones (e.g. R-410A, R-32) in order to comply with the system goals of sustainability (Bhatti et al., 2024).

4.5 Case Studies on HVAC Energy Efficiency

HVAC systems are essential to building systems that provide inhabitants with indoor air quality and thermal comfort. With or without performance loss, energy-efficient HVAC technologies are designed to maximize energy use, operating expenses, and environmental discharge. In this section, we cover the popularity and significance of these technologies, and their definition as well as the categories they belong to, along with their overview.

The energy-efficient HVAC technologies comprise devices, systems, and techniques intended to minimize energy waste and consumption in the heating, cooling, and ventilation operations. Since buildings account for a significant portion of the world's energy consumption and greenhouse gas emissions, these are essential for sustainable building practices. Improve building occupant comfort, productivity, and operating cost and environmental footprint through a reduction in the footprint of buildings through improved efficiency of the HVAC system.

Heating systems that consume less energy try to heat in a warm way. Examples include:

- **High-efficiency furnaces:** Conversion of more fuel to heat that decreases energy waste.
- **Heat pumps:** Devices that use less power than conventional heating techniques to heat indoor rooms by transferring heat from the ground or air.

The energy-efficient cooling systems are made to provide a comfortable indoor temperature while using the least amount of electricity possible. A few examples are:

- **Electricity-efficient air conditioners:** These machines use cutting-edge technology, such variable speed compressors, to use as little electricity as possible
- **Evaporative coolers:** These energy-efficient air-cooling devices use water evaporation rather than traditional air conditioners.

A review of energy-efficient ventilation systems that produce little energy loss while returning fresh air. Among the examples are:

- **Energy recovery ventilators (ERVs):** These devices exchange stale air from indoor or outdoor spaces for new air that has been heated or cooled.
- **Demand-controlled ventilation (DCV):** These systems modify ventilation rates based on air quality and occupancy levels to prevent energy waste.

These technologies optimize the operation of HVAC systems by utilizing sensors, algorithms, and automation. Among the examples are:

- **Programmable thermostats:** These are devices that reduce energy use while a room is not in use and set the temperature to a specific schedule type.
- Building energy management systems, or BEMS, are devices that keep an eye on and regulate lighting, HVAC, and other building systems to ensure maximum energy efficiency.

When combined, these types of energy-efficient HVAC systems help to lower energy use, improve building efficiency, and create extremely sustainable built environments (Tosin Michael Olatunde et al., 2024).

Benefits of Energy-Efficient HVAC

There are many benefits in improving energy efficiency in a building and this includes all its occupants and the surrounding environment. The key advantage of these technologies is described here, including energy savings, better indoor air quality, sustainable environment, and more building resilience.

The potential for significant energy and cost savings is the most evident advantage of energy-efficient or sophisticated HVAC solutions. Lowers utility bills because the electricity and fuel consumption of buildings is kept down by the use of the most advanced technologies and strategies which will help them to use energy optimally. Energy-efficient HVAC systems can save money over time by extending the lifespan of equipment and lowering maintenance costs.

High-efficiency HVAC systems provide for improved temperature, humidity, and ventilation control, which can improve indoor air quality and occupant comfort. With high-efficiency filters and ventilation systems, pollutants and allergens can be eliminated from the air in a healthier setting. In addition, they also come with advanced controls and zoning system that enables the conditions of various part of building to be customized specifically for occupants.

Energy efficient HVAC technology also helps make major contributions in environmental sustainability and the reduction of carbon footprint. These kinds of technology try to lower energy usage, which lowers greenhouse gas emissions and slows down climate change. As a result, energy-efficient HVAC systems can also encourage the adoption of renewable energy sources, further mitigating their negative effects on the environment.

By enhancing system performance and decreasing dependency on external energy sources, HVAC energy-efficient technology will further improve building resilience and reliability. For example, smart controls and automation can be included in buildings to help them adjust to changing circumstances and maintain comfort during severe weather events or power outages. The overall reliability of HVAC systems is also increased by the fact that energy-efficient HVAC systems are often made to last longer and have fewer malfunctions (Tosin Michael Olatunde et al., 2024).

4.5.1 Retrofitting HVAC in Commercial Buildings for Energy Savings

Commercial buildings' largest energy consuming systems are heating, ventilation, and air conditioning (HVAC) systems which can use up to 40% to 60% of the total energy used. And since energy costs continue to rise and sustainability is becoming an issue, retrofitting of current air conditioning systems to improve energy efficiency, lower costs and diminish environmental impacts is becoming an option.

HVAC systems also operate to regulate air quality, humidity levels, as well as occupant comfort in commercial buildings — not just temperatures — making it essential for buildings to be outfitted and taken care of, if needed, by HVAC companies. But a lot of buildings that are still standing use old and inefficient HVAC systems which contribute to astronomical energy usage, much excess maintenance costs and environmental stress.

As technology for HVAC systems advances rapidly and regulators calling for businesses to cut their carbon footprint, more and more organizations are keen to upgrade their systems.

When retrofitting an HVAC system, one has a strategic balance of performance improvement and cost effectiveness. Retrofitting offers opportunity to businesses to integrate modern technology and efficiency improvements with minimal disruption or expense, as it can be done without full system replacement. HVAC retrofits offer an actual route to the energy conservation and sustainability goals by upgrading components, using smart controls, or integrating of renewable energy sources.

Retrofitting involves upgrading or modifying an existing HVAC system to improve upon them without replacing them all together. In this process, we can integrate modern technologies, optimate system controls, and improve insulation in order to improve efficiency and consume less energy. While retrofitting allow for limited business disruption without being quite as cost effective as full system replacements, having a well-planned strategy makes all the difference on your decision (Obinna Iwuanyanwu et al., 2024).

Benefits of HVAC Retrofitting

Retrofitting can save your energy consumer and therefore reduce utility bills. VRF or high efficiency heat pump systems offer advanced sophistication that provides exact temperature control and also, optimizes energy use.

- **Improved Indoor Air Quality:** In addition, the enhanced filtration and ventilation solutions remove the contaminants and provide a healthier indoor environment for the occupants of such buildings.
- **Extended Equipment Lifespan:** A significant value of an HVAC system upgrade is to extend parts life for key components

(compressors and controls) thereby delaying costly replacement on these components.

- **Environmental Benefits:** In addition, HVAC retrofits minimize harmful greenhouse gas emissions that support corporate sustainability goals by decreasing energy usage.
- **Regulatory Compliance:** A majority of governments and local government comprise the energy efficiency standards. Retrofitting helps these businesses comply with these regulations and may entitle them to such incentives and rebates (Ministry of Energy Transition, 2021).

Key Strategies for HVAC Retrofitting

- **Smart Controls and Automation:** Programmable thermostats, occupancy sensors, and so on can optimize the time of heating and cooling to diminish unneeded energy use.
- **Upgrading to High-Efficiency Equipment:** Replacing old compressors, chillers and air handing units with energy efficient units may be worthwhile in savings.
- **Enhancing Building Insulation and Sealing:** Insulating walls, windows, and ductwork minimizes the amount of energy loss, which makes it easier for the HVAC system to operate.
- **Integrating Renewable Energy Sources:** Solar assisted HVAC systems and geothermal heat pumps make the use of sustainable heating and cooling solution minimizing the use of traditional energy sources.
- **Regular Maintenance and Performance Monitoring:** It also includes routine inspections, filter and coil cleaning, and monitoring system performance through watching out through the IoT enabled devices to be sure that such device runs optimally and does not waste energy.

4.5.2   HVAC Energy Optimization in Data Centres

The global reliance upon digital infrastructure, data centres have become today's most consequential facilities in terms of square footings and consume and gobble up energy in ways that are more energy intensive than any other facility in modern industry. At the same time that organizations are processing huge amount of data, supporting cloud computing, powering artificial intelligence and enabling real time digital transactions, the energy demand from these facilities has skyrocketed. The cooling systems of data centre are among the various operational requirements of data centre, especially that they are very important for the operation of data center in order to ensure proper work, continuity of operations, preventing hardware failures to ensure uninterrupted service delivery. However, these classical cooling methods generally use too much energy which increases both the energy operating cost and environmental impact.

Data centres use a lot of energy, and the HVAC system—which accounts for up to 40% of total power consumption—is frequently the most used component of the infrastructure. As with other technologies, inefficiencies in HVAC operations lead to high energy bills and higher carbon footprint which makes cooling solutions in the data center sustainable and that further optimized is a top need of the data center operators. Servers and networking equipment suffer performance degradation, reduce lifetime, and even cause system downtime if not properly temperature and humidity controlled causing high cost in financial losses (Asgari et al., 2021).

In order to overcome these challenges, HVAC energy optimization has become a focus area for data centre management capable of reducing energy waste while preserving the optimal cooling efficiency. A lot of these advancements in technology, including artificial intelligence (AI) based automation, predictive analytics, liquid cooling, economization techniques, renewable energy integration, can provide substantial reductions in energy efficiency and reductions in the environmental burden assigned to the organization.

It presents the fundamental principles of HVAC energy optimization by exploring the fundamental principles of HVAC cooling strategies, best practices, and the latest innovations for the improvement of data centre energy efficiency. In addition, it also points out the advantages of the economic and environmental of the energy efficient cooling solutions as well its insight to the future trends that shape the data centre cooling. The demand for high performance computing is growing, and sustainable and intelligent cooling strategy are a must need, resulting in the evolution of data centres towards the energy efficient and more eco-friendly future.

It is common knowledge that data centers produce many tons of heat through continuous server operation, storage system operation and networking devices operation. In order to maintain performance, avoid equipment failures, and comply with ASHRAE (American Society of Heating, Refrigerating, and Air Conditioning Engineers) regulations, temperature and humidity must be regulated. Nevertheless, conventional cooling methods cause inefficiency, consume large amount of power, and come at high cost of operation (Steinbrecher & Schmidt, 2011).

The Energy Optimization Key Strategies for HVAC

- **Precision Cooling and Airflow Management:** Traditionally, cooling is achieved through processes that often over cool and suspend with unsuitable wind distribution. Hot aisle/cold aisle containment and underfloor air distribution as containment strategies make airflow financially efficient, reducing energy waste. Additional optimization of hotspots and lost values can also be accomplished using Computational Fluid Dynamics (CFD) modeling.

- **Utilization of Economization Techniques:** Economization or free cooling is based on external ambient air or water sources that reduce the amount of mechanical cooling needed. Economizer loops are dedicated water side loops or dedicated air side loops, and are brought in as a means of using the cool ambient air outside, when such conditions are found to be favorable. These methods significantly lower energy consumption, especially in regions with colder climates.

- **AI-Driven and Predictive HVAC Optimization:** These efforts that integrate artificial and machine intelligence to effectively monitor and predictively maintain HVAC systems are now feasible. They are smart sensors and iot enabled controls that dynamically adjust cooling outputs to match load fluctuations and but cannot save unnecessary power usage and can offer the best thermal conditions. Additionally, AI driven analytics can also predict the system failures so down time and maintenance cost can be minimized.

- **Liquid Cooling Technologies:** Traditional air cooling is still the dominant way but as direct to the chip or immersion cooling gain in popularity due to their excellent thermal efficiency. These technologies directly transfer heat into a liquid medium and in so doing lower HVAC loads, improve cooling efficiency and support higher density computing environments.

- **Renewable Energy Integration:** Reducing carbon footprints and reducing dependency on grid electricity are two benefits of incorporating renewable energy sources like solar and wind power into HVAC lots. Those coupled with hybrid cooling based on renewables and renewables, energy resilience and sustainability in data centre operations (Nwagu et al., 2025).

4.5.3 Case Study: Achieving Net Zero Energy in Office Buildings

The change concerns intensifying and energy costs on the rise, businesses and their respective governments around the world are moving in a direction towards sustainable building solutions. A relatively large energy efficiency improvement opportunity exists in office buildings, which represent a major percentage of the energy use in cities. Net Zero Energy (NZE) buildings are gaining its popularity as a recourse to minimize the environmental environment while let operational efficiency reach its set limit. An office block is considered to be Net Zero Energy if it generates as much energy as it uses in a year, usually by combining smart energy management technologies, renewable energy sources, and superior energy efficiency.

The case that is presented in this study explores the successful absorption of a real-life office building into the category of (NZE) buildings. It features how they did things, obstacles they faced, and what key learning points can be delivered to other projects in this field or industry. This case study analyses to what extent NZE buildings are practically feasible for the commercial sector and if NZE buildings bring about environmental, financial and operational benefits (Jaysawal et al., 2022).

The key objectives of the project were:

- Innovation in terms of the architectural design to reduce the overall energy consumption, the use of high-performance HVAC systems and improved lighting solutions.
- Change the structure of the given sentence.
- With smart building technologies, such as automated lighting, air quality monitoring as well as dynamic temperature control; smart building technologies raises occupant comfort and productivity.
- Reduces energy expenses in the long term thus achieving cost savings and wins government incentives for sustainable buildings.

Key Strategies Implemented

1. Energy-Efficient Building Design and Retrofitting

- The high-performance insulation coupled with low emissivity glass windows, plus an optimized building envelope was used to minimize heat loss and gain through the building.
- Strategic building orientation, natural ventilation, daylight harvesting, all of which reduced reliance on artificial lighting and HVAC systems, were some of the passive design strategies.

- LED lighting with motion sensors and daylight dimming decreased electricity consumption by a great margin.

2. Smart HVAC and Energy Management Systems

- A high efficiency HVAC system comprised of variable refrigerant flow (VRF) technology, demand-controlled ventilation and heat recovery systems were included in the building.
- Real time energy usage was monitored by smart thermostats, or IoT enabled energy management systems that automatically adjusted heating, cooling, and ventilation in accordance with occupancies, external weather conditions among other things.
- Similarly, AI driven predictive maintenance was integrated to ensure that HVAC equipment is maintained at peak efficiency so that energy wastes are eliminated by malfunctions or inefficiencies.

3. Renewable Energy Integration

- The clean electricity is generated annually through on site solar photovoltaic (PV) panels be installed on the building roof and also on adjacent parking lot.
- An excess solar energy storage system was used to store excess solar energy for use during the night and peak demand periods.
- The geothermal heating and cooling systems utilized the underground thermal energy in order to reduce the HVAC energy demand.
- If wind turbines were utilized, a further offset occurred with renewable power due to grid dependency.

4. Water and Resource Efficiency Measures

- Rainwater harvesting systems collected and filtered in order to use them for irrigation and restrooms.
- Greywater recycling systems reduced water waste in a coordinated way towards the overall sustainability goals.

- Water efficient landscaping along with low-flow plumbing fixtures secured the water savings against Net Zero Water principles.

6. **Behavioural and Policy Changes**

- To inspire energy conscious behaviour, sales forces and employees received awareness campaigns along with energy usage dashboards displayed in common areas in real time and directly affected them.
- Reduction in the total building energy footprint in the office building was facilitated by the flexible work policies, remote work options and even shared workspaces.
- The building received LEED Platinum certification and WELL Building Standard certification and signals the company's continued commitment to sustainability and occupant wellbeing (Kinetik, 2011).

Currently, the Net Zero Energy model, although technically unrealistic, is achievable for office buildings around the world as technology and energy policies develop. Those who invest proactively in sustainable building solutions will decrease their operational costs, build strong corporate reputation as well as help the world achieve global carbon neutrality.

4.6 Chapter Synthesis

In this chapter, all that can be found about HVAC energy efficiency from performing energy audits, to performance monitoring, to predictive maintenance and real world case studies, became covered. The importance of HVAC energy audits is then explained, and it describes methodologies of assessing system performance, identifying inefficiencies and making recommendations for improving energy savings. First, audits as a foundational step are upon which the optimization of HVAC operations is based, discovering operational deficiencies and areas for potential gains in efficiency.

In the end it gets into performance monitoring and predictive maintenance, and makes explicit why data analytics plays such an integral role in HVAC energy management. Data analytics helps in optimizing energy consumption in terms of minimizing energy expenditure as well as lower operational costs by making use of trend analysis and predictive modeling within the real time. These also include fault detection and diagnostics (FDD), that is, identification of anomalies and prevention of system failure before failure has occurred. Preventive vs. predictive maintenance comparison is done and advantages of predictive techniques are weighed, namely the reduction of downtime, their cost effective apply and their ß-advantage of equipment 's lifespan.

The chapter then goes on to investigate retrofitting and upgrading HVAC systems in terms of technological advances in high efficiency equipment, smart controls and automation towards improved energy performance. A large contribution of reducing carbon footprints as well as achieving a sustainability goal is through retrofitting.

Real world case studies of how HVAC energy efficiency initiatives affect the business are finally presented. Such approaches include retrofitting HVAC systems in commercial buildings, boosting energy efficiency of data centres, and a case study on how to attain zero energy buildings by means of integrated HVAC strategies. However, these studies also reveal the effectiveness of data driven decision-making and improvements of the system in attaining sustainable energy efficiency. This chapter integrates audits, advanced analytics, and innovative technologies to structure a prescription for addressing the energy issues associated with HVAC operations, and provides assurance for achieving a long-term savings.

Multiple Choice Questions (MCQs)

1. **What is the primary objective of conducting an HVAC energy audit?**

   a. To increase HVAC operational costs
   b. To identify energy-saving opportunities and inefficiencies
   c. To replace all existing HVAC components
   d. To eliminate the need for regular maintenance

2. **Which technology plays a crucial role in HVAC energy management?**

   a. Manual record-keeping
   b. Data analytics and IoT sensors
   c. Incandescent lighting systems
   d. Paper-based maintenance logs

3. **What is the main advantage of predictive maintenance in HVAC systems?**

   a. It allows for emergency repairs only
   b. It prevents potential failures before they occur
   c. It is more expensive than reactive maintenance
   d. It reduces the need for regular inspections

4. **Which technique is commonly used in fault detection and diagnostics (FDD) for HVAC equipment?**

   a. Guesswork by maintenance staff
   b. Real-time data monitoring and machine learning algorithms
   c. Manual thermostat adjustments
   d. Increasing cooling setpoints randomly

5. **How does retrofitting an HVAC system improve energy efficiency?**

    a. By replacing inefficient components with energy-efficient alternatives
    b. By increasing overall energy consumption
    c. By eliminating the need for temperature control
    d. By reducing air filtration capabilities

6. **Which of the following is a common retrofit strategy for improving HVAC efficiency?**

    a. Installing high-efficiency chillers and variable speed drives
    b. Keeping outdated HVAC units in operation indefinitely
    c. Disabling temperature controls to reduce energy consumption
    d. Increasing air leakage in ductwork

7. **What is a key consideration when optimizing HVAC energy use in data centers?**

    a. Maintaining excessive cooling for all equipment
    b. Implementing airflow management and precision cooling
    c. Ignoring energy consumption patterns
    d. Reducing server loads by turning off HVAC systems

8. **In achieving net zero energy in office buildings, what is a crucial factor?**

    a. Maximizing energy waste for better airflow
    b. Integrating renewable energy sources with efficient HVAC systems
    c. Using outdated refrigerants to cut costs
    d. Avoiding any building insulation improvements

9. **Which of the following is a primary goal of an HVAC energy optimization case study?**

   a. To demonstrate successful energy-saving strategies
   b. To prove that HVAC efficiency improvements are unnecessary
   c. To increase energy consumption for comfort
   d. To eliminate HVAC systems altogether

10. **What is a key takeaway from HVAC case studies on energy efficiency?**

    a. Retrofitting and optimization strategies can significantly reduce energy costs
    b. HVAC upgrades have no impact on energy consumption
    c. Energy-efficient HVAC systems always require more maintenance
    d. Older HVAC systems are inherently more efficient than newer models

Answers

1	2	3	4	5	6	7	8	9	10
c	b	b	a	d	a	b	b	d	a

CHAPTER 05

FUTURE TRENDS, BEST PRACTICES, AND CONCLUSION

5.1 Chapter Overview

It is dedicated to the technologies emerging in energy efficiency of HVAC systems, discussing how these new innovations such as digital twins, AI, and blockchain are adjusting HVAC systems to reduce energy conditions along with raising efficiency. The first examines the digital twin and virtual commissioning use for real time monitoring and simulation that allow engineers to optimize HVAC configurations prior to installation. Its predictive maintenance and automated climate control are especially being carried out in smart city infrastructure with the help of AI driven HVAC systems. By using blockchain technology HVAC energy management is more transparent and secure, enables peer to peer energy trading as well as tracking of the data. Finally, the chapter also provides best practices for the implementation of these technologies for engineers and facility managers to facilitate maximum energy efficiency and longevity of the system. This concludes with key strategies for integrating this sort of advancement into low carbon building operations and developing the future of HVAC in sustainable environments.

5.2 Emerging Technologies in HVAC Energy Efficiency

Reductions in energy consumption on building in heating, ventilation, and air conditioning (HVAC) systems are taken to the forefront of the field. Due to the fact that buildings use a large portion of global energy, there is a great need for more efficient HVAC solutions. Smart, more responsive systems currently emerging are providing a transformative role by allowing these systems to adapt to environments with changing conditions and the needs of the user. digital twin and virtual commissioning are the ways an engineer can make a detailed simulation of an HVAC system that can be monitored and then optimized in real time. An artificial intelligence (AI) serves to automate the system and also helps in predictive maintenance which interns reduces the downtime and improves overall energy efficiency. At the same time, blockchain technology is being used to secure the energy data, enable peer to peer energy trading and ensure energy usage transparency(O'Keefe et al., 2010).

The technologies themselves are being integrated to smart, sustainable HVAC systems. These artificial systems can study massive inputs, thereby predicting the heating and cooling requirements, and automatically change the system settings to minimize the loss of energy. Smart thermostats and the advanced climate control systems let the users personalize the climate settings and also optimize the energy use. The energy transactions form a good opportunity for energy management and saving cost when done on the blockchain: it's secure and traceable. In order for engineers and facility managers to adopt these new technologies, a strategic approach must be taken into account that includes system design, operational efficiency and long-term sustainability. This chapter studies how these innovations are changing HVAC in the future in terms of best practices and future trends of energy efficient building management.

The Latest Emerging HVAC Technologies

The demand for energy efficiency and personal comfort is increasing, which means the innovation on how heating, ventilation and air conditioning systems operate is also going up. Apart from lowering the energy consumption, these advancements also tend to increase user experience and make systems more reliable(Reddy et al., 2016).

Smart Thermostats: Personalized Climate Control

One of the key advancements in emerging HVAC technologies is the development of smart thermostats. These innovative devices are designed to provide personalized climate control, allowing homeowners to optimize their comfort while minimizing energy consumption. Unlike traditional thermostats that require manual adjustments, smart thermostats use advanced sensors and algorithms to learn your preferences and automatically adjust temperature settings accordingly.

Figure 5.1: Smart Home Thermostat Control.

Source: (Pro, 2024)

With a smart thermostat, you can easily create customized schedules that align with your daily routine. For example, you can program the thermostat to lower the temperature while you're away at work and raise it just before you return home, ensuring a comfortable environment upon your arrival. Some models even have geofencing capabilities, which means they can detect when you're approaching home and start adjusting the temperature accordingly.

But what truly sets smart thermostats apart is their ability to connect with other smart home devices. By integrating with platforms like Amazon Alexa or Google Assistant, you can control your thermostat using voice commands or through smartphone apps. This level of convenience allows for seamless connectivity between different aspects of your home automation system.

Energy-Efficient HVAC Systems: Redefining Sustainability

In today's world, energy efficiency is a top priority for homeowners and businesses alike. Emerging HVAC technologies are addressing this concern by introducing energy-efficient systems that not only reduce environmental impact but also help save on utility bills.

New HVAC systems incorporate advanced components such as variable-speed compressors and high-efficiency heat exchangers that optimize performance while minimizing energy consumption. These systems are designed to operate at different speeds depending on the heating or cooling demand, ensuring that they don't consume unnecessary energy when it's not needed.

Figure 5.2: HVAC Technician Repairing Air Conditioner.

Source: (Pro, 2024)

Additionally, many energy-efficient HVAC systems come equipped with smart features like occupancy sensors and adaptive controls. These features allow the system to adjust its operation based on occupancy patterns, ensuring that energy is not wasted in unoccupied spaces. For example, if a room is empty for an extended period, the system can automatically reduce heating or cooling output to conserve energy.

The Rise of Zoning Technology: Customized Comfort for Every Room

Those days when there was only one thermostat controlling the temperature of an entire home are gone. With the uprise of zoning technology, homeowners have the chance to get the comfort of their own room. Zoning systems separate a home into separate zones with their own thermostats and controls. It lets you set different temperatures for different areas in the house based on how people in that area prefer and use the place. In other words, bedrooms can be cooled at night and common areas in the day without being too cold. Zoning technology guarantees that all members of the household have comfort levels at the level they desire without sacrificing energy efficiency.

Also, zoning systems may allocate certain areas before others. When a room demands more heating or cooling because of windows or direct sunlight exposure, you can use extra resources (heat or air conditioning) to that particular zone, without affecting the other rooms of the home. Zoning tech increases comfort and energy efficiency to that extent, so it's a worthwhile improvement in modern HVAC systems.

Air Purification Innovations: Breathing Fresh, Clean Air

Indoor air quality is a crucial aspect of comfort and well-being. Emerging HVAC technologies are incorporating air purification innovations to ensure that you breathe fresh and clean air within your living spaces. Many modern HVAC systems come equipped with advanced filtration systems that capture dust particles, allergens, and even harmful pollutants from the air. These filters are designed to trap microscopic particles and prevent them from circulating throughout your home. Some high-end systems even use UV-C light technology to neutralize bacteria and viruses present in the air.

Furthermore, some HVAC systems have the ability to monitor and analyze air quality in real-time. They can detect pollutants such as volatile organic compounds (VOCs) and adjust ventilation rates accordingly to maintain optimal air quality. This feature is particularly beneficial for individuals with respiratory conditions or allergies.

Integration with Smart Home Devices: Seamless Connectivity

As mentioned earlier, emerging HVAC technologies are designed to seamlessly integrate with other smart home devices. This integration allows for centralized control and enhanced automation capabilities.

By connecting your HVAC system to a smart home hub or using voice commands through virtual assistants, you can effortlessly manage your comfort settings from anywhere in your home. For example, you can adjust the temperature, change fan speeds, or even turn on/off the system using a smartphone app or by simply speaking a command.

Figure 5.3: Smart Thermostat in a Modern Home.

Source: (Pro, 2024)

The integration of HVAC systems with other smart devices also opens up opportunities for advanced automation. You can create customized routines that trigger specific actions based on different scenarios. For instance, you can set up a routine that automatically adjusts the temperature when you leave home or turns off the system when no one is present in the house.

5.2.1   Digital Twins and Virtual Commissioning in HVAC

The digital twins and the virtual commissioning integration is bringing a great change in the HVAC industry making it better by enhancement in system design, testing and operational efficiency. By coupling these technologies with advanced modeling, simulation and optimization of HVAC systems prior to their installation, errors associated with building specific HVAC energies and operating schedules are decreased, cost is minimized and overall system performance is improved(Shyam Varan Nath, Pieter van Schalkwyk, 2021).

A digital twin, as the name suggests, is a virtual representation of a physical HVAC system that is running real time. Then in this virtual model is based on the physical system with sensor reading, equipment specification, and environmental factors. By continuously monitoring the system performance, digital twin lets the engineer and facility manager predict the potential failure and optimize the efficiency. Machine learning algorithms enable the data to be put to use in identifying patterns to forecast maintenance needs and set system settings to avoid breakdowns. Moreover, engineers can run various configuration and control strategies within the digital twin and test them before actually putting them in the physical system, thus minimizing trial and error on the field. Digital twin is a good example, such as a large commercial building facility manager can change airflow or different types of filters while the digital twin simulates on adjustment energy consumption and air quality for the facility manager to learn without affecting routine facility operations.

Virtual commissioning is the use of simulation, virtual modeling and testing and configuration of an HVAC system prior to physical installation. Typically, traditional commissioning processes rely on extensive on-site testing, adjustments, time, and cost. Virtual commissioning tackles these problems by providing engineers with the ability to test the total control logic and automation process within a virtual environment to find, and fix, errors prior to deployment. In addition, it certifies that everything works as it should, including sensors, actuators and controllers. The commissioning time on site is greatly reduced by resolving problems in the virtual model, which reduces labor costs and disruptions. Virtual commissioning, for example, is used to test different ventilation rates, thermostat settings and zone configurations prior to installation when, for example, an HVAC system is to be installed in a high-rise building.

Figure 5.4: Smart Home Control on a Tablet.

Source: (Pro, 2024)

Digital twins and virtual commissioning have a lot of benefits in HVAC. It is constantly adjusted and monitored to ensure that energy use is maximized and efficiency in the whole is optimized. Predictive maintenance is done by helping of machine learning and real time data analysis to stretch out the life of the equipment and keep the failures expected. Early detection of the problem and Realtime adjustments in the system reliability and when the zoning and airflow is adjusted dynamically the temperature and air quality is in the same range. Pre testing and virtual modeling also helps to minimize installation cost and labor expenses, thereby minimizing overall commissioning costs. The use of digital twins and virtual commissioning is now setting the standards for the design and management of the HVAC system. The ways combine the real time data, prediction model as well as automation to enable HVAC related operations to be more efficient, cost effective and reliable.

5.2.2   AI-Driven HVAC Systems for Smart Cities

As smart cities rise, AI introduced HVAC systems are becoming the new norm in their adoption to make HVAC systems more energy efficient, sustainable, and occupant comfortable. AI technology empowers the HVAC system in smart cities to dynamically adapt according to the changes in the environmental conditions, such as temperatures, and energy demands as well as the user preference. AI driven HVAC systems take the benefit of real time data from linked sensors, weather prediction, and building occupancy pattern, in order to optimize HVAC operations to reduce energy use and environmental effect(Mishra & Singh, 2023).

Large scale data collection and worrying is vital for enhancing the performance of smart cities' AI driven HVAC systems. Building and city infrastructure is equipped with sensors that feed data on the temperature, humidity, CO_2 levels, weather conditions as well as occupancy. This data is machine learning algorithms to or looks for patterns and predicts the future condition. Let's consider an example of an AI driven HVAC system in a smart office building that detects the missing of occupancy and the HVAC output accordingly can be adjusted to save the energy at the same time comforted. Similarly, integrating in weather forecasts can be utilized in the AI model to anticipate changing outdoor temperatures and the system can disseminate preseason components of comfort inside temperature to ceaselessly keep up with indoor settings.

The major benefit of AI driven HVAC systems is that they can optimize energy consumption at city wide scale. In a smart city, several buildings, and infrastructure components are connected to a central management system. The AI algorithms monitor energy patterns across the city and adjust the HVAC operation according to the grid demand and availability of energy. For example, the system can lower cooling intensity in non-critical areas whilst concurrently maintaining a comfortable condition in high priority areas. A load balancing capability that prevents grid overload and enhances more stable distribution of energy across the city is one of this load balancing capability.

The predictive maintenance and self-optimization of the system can also be done using AI. The use of AI algorithms to analyze historical performance data can detect early showing of equipment degradation or failure, prior to the time equipment experiences a breakdown. The proposed approach reduces downtime and increases the life span of HVAC equipment. Additionally, AI driven systems can change and learn with change in building use or environmental condition. For instance, if a comparable building is erected nearby, then changing wind patterns or shading may, in turn, affect the efficiency of the HVAC system. Such changes can be detected by an AI driven system and automatically recalibrate airflow and temperature settings to keep on performing at their best.

AI driven HVAC systems in smart cities offer a lot of environmental benefits. AI driven systems optimize energy use saving and cutting back on the amount of waste which helps to offset climate change and achieve sustainability goals within the city. Also, better air quality and temperature control makes for healthier and better life for city people. Smart zoning allows different parts of city to use different temperature settings depending on the usage pattern, use less energy and still keep the occupant comfortable.

However, AI driven HVAC systems are what will establish the smart cities of the future as more sustainable and efficient. These systems are following adaptive control, real time data analysis, predictive maintenance to increase energy efficiency, decrease cost and enhance the comfort and health of city residents. With the evolution of AI technology, AI driven HVAC systems will play more significant part in the building of smarter greener urban environments in smart cities.

5.2.3   Role of Blockchain in HVAC Energy Management

The HVAC energy management is one of the techs which is regarded as a game changer and this is because of the emerging blockchain technology that is being used as a framework to track and optimize energy consumption in a method that is secure, transparent, and decentralized. The conventional method for running HVAC systems is through centralized data storage, management, and security which are prone to inefficiencies, inaccuracies and security breaches. However, the problems are solved by the blockchain because every transaction, as well as energy and any other pattern, is recorded and securely stored in the distributed ledger on several nodes. Since the data is decentralized, it enhances data integrity such that risk of single points of failure or tampering is eliminated(Trout & Betts, 1988).

Smart contracts, or self-executing contracts that have terms embedded in the code, are one of the important advantages of blockchain in HVAC energy management. Indeed, smart contracts can even regulate the settings in HVAC in real time based on real time data and pre-defined rules. For example, in conjunction with energy price spikes, the HVAC system can be called to alter temperature settings to spend less and still be comfortable. Smart contracts can also balance energy consumption at multiple HVAC systems within a building or a smart city to optimize load balancing and decrease wastage of energy.

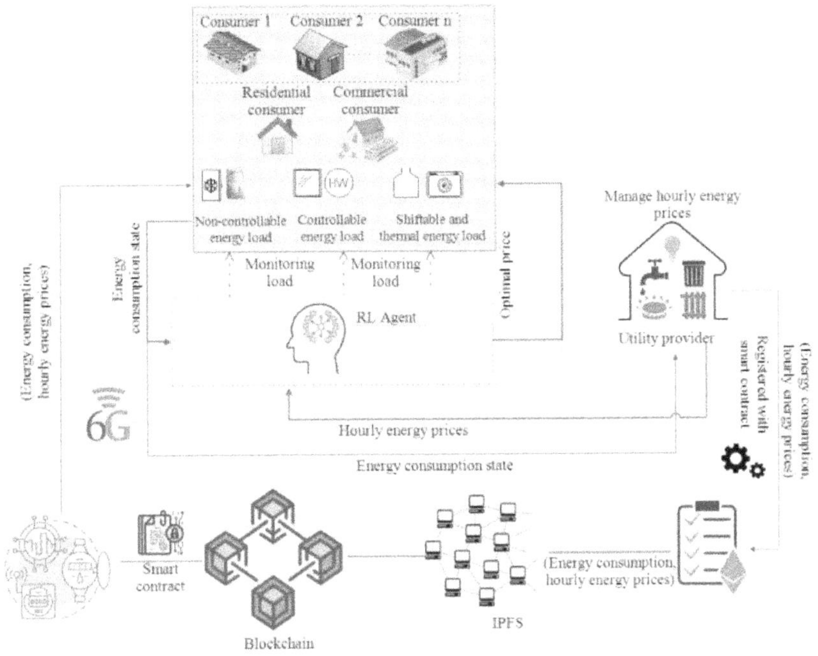

Figure 5.5: AI-driven energy management system for smart grids using RL, blockchain.

Source: (Kumari et al., 2023)

Also, blockchain makes possible peer to peer (P2P) energy trading within smart buildings and micro grids. By tracking and trading excess energy from renewable sources like solar panels directly with other consumers in the network, blockchain based marketplaces allow excess energy to be generated. Such reduction depends on traditional providers of energy and promotes energy autonomy. With blockchain records, all transactions are transparent and verifiable, thus the trust among participants is increased, and fair pricing is possible.

Besides these benefits, blockchain offers an integration with AI and machine learning to improve HVAC system maintenance and fault detection. AI models can, by historical & real time data sampled on the blockchain, analyse patterns, forecast system failure and recommend maintenance action. It provides an ability to predict HVAC operations and this capability mitigates downtime, avoids expensive repairs and contributes more reliable and efficient HVAC operations. Blockchain provides much greater data privacy and security. By the fact that data is encrypted and spread across a network, unauthorized access and data breaches are less likely. They (data providers) can also provide accurate, real time data, as well as help building managers and energy providers to make informed decisions which are also in compliance with data protection regulations(Ucar et al., 2024).

As the way of life turns the direction of smart city and connected infrastructures, but as HVAC systems are getting increasingly interconnected, blockchain presents in the scalable and secure basis to handle the flow between complex energy ecosystems. Using such things as blockchain, AI, IoT, and renewable energy sources to combine them with the HVAC systems in order for them to operate more autonomously, efficiently and sustainably. Unlike conventional approaches, this transformative one does more than reduce operational costs; it facilitates global energy efficiency and carbon reduction goal. This is the next step towards a veritable, adaptive, and green future through the role of Blockchain in HVAC energy management.

5.3 Best Practices for Engineers and Facility Managers

The efficiency, reliability and lifespan of HVAC (Heating, Ventilation, Air Conditioning) system are dependent on engineers and facility manager. Good management not only promotes indoor air quality and comfort but also cuts down the energy in use and its costs of operation. Following are the best practices that provide structured and comprehensive way of maintaining and optimizing HVAC systems that bring immediate and long-term benefits in building performance and occupant satisfaction.

- **Regular Maintenance and Inspection**

Regular maintenance and inspection are the base of efficient HVAC system. The comprehensive preventive maintenance schedule should include all HVAC components such as air handlers, compressors, condensers, chillers, ductwork, and ventilation system. Engineers and facility managers should put the schedule in place in order to prevent breakdown of the HVAC system. In order to prepare the system for increased loads during summer and winter which are often times of high energy demand, seasonal inspections should be conducted. Regularity in changing or cleaning of air filters is important to keep the proper air flow and to increase the indoor air quality. Further, one should check refrigerant levels, inspect insulation, and test actuators, thermostats, and sensors to keep the system in good working condition. Regular maintenance does not only help keep your car performing well but also provides you with a break down early enough and keep you from having costly breakdowns through regular maintenance. Thermal imaging and ultrasonic leak detection can be incorporated into advanced diagnostic tools to make maintenance activity even more accurate and efficient.

- **Optimize System Performance**

Fine tuning of the HVAC system to achieve the highest level of efficiency with the least amount of energy consumption is optimization of system performance. Regular calibration of thermostats and sensors should be done to prevent unnecessary heating or cooling of a building and ensure accurate temperature control by engineers. It is also essential to perform the airflow balancing to remove hot or cold spots to make sure they don't exist anywhere in the building and to make sure it is comfortable in all areas around the building. Pressure levels within the ducts can be monitored and if there are leaks or blockages that prevent proper system performance, this can be detected. It is also possible to introduce zoning controls and smart ventilation systems, that would help to direct airflow only where it is necessary. These are both measures that improve occupant comfort, as well as measure that reduce the overall energy consumption. Additionally, upgrading to smart variable air volume (VAV) systems and installing economizers are ways that outside air can be used for free cooling and help increase energy efficiency.

- **Energy Efficiency Improvements**

Reduction in operational costs and reducing the environmental impact of the HVAC systems involves improvement in energy efficiency. As is usual in any upgrading situation, high efficiency HVAC units with variable speed motors and smart controls can cut energy use by 60 percent or more. To allow these systems to adjust automatically with based on how many people are occupying a room and external temperatures, programmable thermostats and smart building management systems can be installed. Wind energy recovery ventilators (ERV) and demand-controlled ventilation (DCV) systems help reduce energy wastage through recycled conditioned air and constant ventilation needs to be adjusted to the loads of the premises. Engineers and facility managers should also consider renewable energy as an option, such as solar powered HVAC systems, to continue the sustainability. Heat recovery systems capture waste heat from exhaust air, and reusing this in preheating the incoming air can increase the total system efficiency. Furthermore, the energy spikes incurred by the existing system caused by power balancing can be limited by integrating AI-based predictive control algorithms, contributing more effectively to load distribution balancing.

- **Air Quality Management**

There is a key responsibility for facility managers to ensure high indoor air quality (IAQ). Poor indoor air quality can cause health problems, diminish productivity and bring discomfort to occupants.

Figure 5.6: HVAC Technician Performing Maintenance on Air Ducts.

Source: (Lubach, 2024)

However, this is easily solved through the use of high efficiency particulate air (HEPA) filters that will capture airborne pollutants, allergens, and contaminants improving overall air quality. In addition to that, it is important to monitor humidity levels as excess humidity may leads to mold growth and decrease the level of comfort, while lack of humidity will likely cause respiratory discomfort and generate static electricity. Ventilation of course is very important to ensure circulation of fresh air, and ventilation rates should be adjusted according to the ASHRAE (American Society of Heating, Refrigerating, and Air Conditioning Engineers) as well. In addition, engineers should take into consideration the use of ultraviolet (UV) air purifiers to neutralize pathogens and improve IAQ. Further improvements of indoor environment can be achieved by an advanced filtration system, such as activated carbon filters, with removal of volatile organic compounds (VOCs) and odors.

- **Data-Driven Maintenance and Monitoring**

All can be improved using data driven insights that leverage existing HVAC system procedures. An analyst can use BAS to monitor the Realtime performance of HVAC and fix issues very quickly. This technique can also be used to predict the failure of equipment, proactively maintain to minimize downtime. FDD tools can identify inefficiency such as excessive energy use or unnatural temperature fluctuations that can be caused by changes or properly recommend corrective measures. Data driven maintenance is a means to maintain the system at peak performance, without increasing energy usage, and further, optimize energy usage and prolong the life of the system. Once these machine learning algorithms are integrated into the BAS systems, learning is continuous while the BAS systems are making changes (or adapting) based on both the real time and historical data. With a cloud-based platform, the HVAC performance metrics can be accessed centrally, to then streamline remote troubleshooting, and also do performance benchmarking between different facilities.

- **Training and Workforce Development**

To continue to provide an educated and skilled HVAC team, training and workforce development investments are the best way to go. Technicians have to receive ongoing training from engineers and facility managers, as the equipment installed into them always changes in compliance with regulatory changes, latest HVAC technologies, and best practices. Such collaboration can foster better problem solving and operational efficiency if encouraged between engineers, maintenance groups and building occupants. Standard operating procedures (SOP) for emergency repairs, system shutdowns and routine maintenance set up an organized way to deal with air conditioning systems. Such a workforce, well trained, is more able to face and solve complex HVAC challenges and to deliver high performance outcomes. To continue to be at a high level it is important for the industry to continue to support certification programs like LEED, EPA section 608 and ASHRAE training. At the same time, the cultivation of a climate of innovation among technicians whereby these employees are solicited to inculcate improvements to the process can yield a culture of continuous operational improvements.

- **Sustainable Practices and Compliance**

The environmental footprint of buildings can be reduced as well as long term efficiency improved by implementing sustainable practices in HVAC management. It is important for engineers to choose environmentally friendly refrigerants like R-32 or R-1234yf that have less global warming potential compared to classic refrigerants. To avoid severe penalties and safe operation, it is necessary to comply with local environmental regulations and industry standards. Achieving higher energy efficiency and enhanced indoor environmental quality can be realized by designing HVAC systems which can satisfy the LEED (Leadership in Energy and Environmental Design) standards. Solar and geothermal heating and cooling are types of renewable energy sources that further enhance sustainability. In fact, facility managers should also look to see how they can reduce waste, like recycling air filters, as well as optimize energy use in non-peak hours. The monitoring and reduction of environmental impact of HVAC operation can be achieved by adopting carbon footprint tracking and reporting systems. Facility can add further to their sustainability by integrating carbon offset programs as well as smart energy purchasing strategies.

- **Emergency Preparedness and Risk Management**

Emergency preparedness plan for HVAC system is critical for keeping HVAC system reliability and occupant safety. Protocols should be developed by engineers and facility managers for responding to system failure, power outage and extreme weather events. Uninterruptible power supplies (UPS) and generators are backup power sources and should be tested regularly to make sure they will support the HVAC functions in the event of emergencies. In general, regular emergency drills, training of staff on proper procedures to respond rapidly in an emergency will minimize downtime and minimize damage in the case of unforeseen events. Therefore, engineers should also set up a system for monitoring weather conditions and adjust HVAC operations in advance to avoid overload or system failure.

5.4   Conclusion and Roadmap for the Future

New technologies, regulatory fixes, and constraints to energy efficiency in the HVAC realm collectively position the future for a massive growth in its pursuit for greener solutions. The role that AI and IoT are playing in optimizing HVAC performance is by providing real-time monitoring, prediction of maintenance, and smart climate control. These smart systems facilitate in reducing energy waste, operational costs and improve comfort and efficiency. In addition, the energy sources from the renewable side, like solar powered HVAC, geothermal heating and cooling, and waste heat recovery, are transforming the sector due to the reduction in the use of fossil fuels as well as overall carbon emissions(Francis, 2004).

At the same time, in materials science, phase change materials and high-performance thermal insulation are improving retention of energy and also efficient HHVC. Simultaneously, electrification is also picking up steam from the moving away from traditional gas heating to electric powered heat pumps that provide superior energy efficiency and a much lower environmental impact. And it's at the same time that governments around the world are increasing tightening of their energy efficiency standards, as well as giving incentives to the green building initiatives, which are forcing businesses and households to go towards HVAC that is environment friendly. Another trend pertains to smart grids integration so that HVAC systems can contribute to demand responsive programs and consume energy as smartly as possible to the grid as a whole as well.

In the future, the industry will continue its short-term focus of increasing adoption of AI driven HVAC systems as well as retrofitting existing older buildings with energy efficient tech. However, in the midterm, renewable powered HVAC solutions will be widespread and solid-state cooling will continue to evolve resulting in further improvement on system efficiency. In the long run, a fully autonomous and AI optimized HVAC system development will allow buildings to actively self-regulate energy use dynamically contributing to a net zero energy buildings trend. The future of the HVAC industry lies in energy modeling with quantum computing and advanced simulations, and advances in energy modeling will become more powerful with the result of how the HVAC industry will continue to evolve a future where sustainability and efficiency are occurring hand in hand.

In this world changing towards the carbon neutrality, the HVAC industry needs to develop innovation and collaborate to meet the increasing demand for the energy efficient solutions. In order to minimize the consumption of energy and the impact on the environment, AI, renewable energy and smart building technologies will have a major role to play. The adoption of these advancements will provide the businesses and the policymakers a way to pave the path for a sustainable future in HVAC energy efficiency, creating long-term benefits not only for the industry but also for the planet.

5.4.1   Summary of Key Energy Efficiency Strategies

Energy efficiency in the HVAC systems is the result of advanced technologies, smart management practices and sustainable design principles. Hence, a variety of strategies are being explored to improve the performance of HVAC in energy efficient buildings with less environmental impact. An integrated approach is to integrate AI and IoT smart controls for smart controls with real time monitoring, predictive maintenance and adapting climate control. With such intelligent systems, occupancy patterns, external weather conditions and indoor air quality can be analyzed to adapt HVAC control operations, so as to obtain optimal performance, while significantly reducing the energy waste. Also, automated demand response enables an HVAC system to interact with electricity grids to adjust cooling and heating loads according to peak or off-peak hours. It enables reduction of energy costs, strengthening of grid stability, and supporting the broader pathway to sustainable energy consumption.

Figure 5.7: Solar-Assisted Heat Pump on a Rooftop.

Source: (Editorial, 2023)

Adopting renewable energy sources, like solar powering HVACs, geothermal heating and cooling, as well as waste heat recovery systems are also deeply important. Renewable solutions ensure these provide a partial solution to energy dependence on fossil fuels and provide long-term energy savings. Later, high efficiency HVAC equipment such as; variable refrigerant flow (VRF) systems, energy recovery ventilators (ERVs) and magnetic bearing chillers, has further aided in the industry progress by producing optimum performance with much lower energy consumption. Furthermore, by employing advanced insulation materials, high performance windows, and phase change materials, the heating and cooling loads can also be well reduced and the overall efficiency of HVAC systems improved. Passive design strategies enhance active energy saving measures and are important to minimizing the amount of energy needed at the building level.

Moreover, the advent of the electric heat pump as a mainstay HVAC system is gaining traction amongst major climate change suppositions like a significant approach to boost energy efficiency and decrease carbon emissions. Furthermore, the adoption of district heating and cooling networks further reduces the cost of thermal energy by distributing thermal energy across a number of buildings. Introducing smart grids and building automation systems together with the ability to integrate the HVAC units with other building system to dynamically optimize the energy usage is another significant advancement. The applications of these technologies include demand side energy management to reduce peak demand load, as well as efficient operation of renewable energy sources. Maintenance, energy audit, system retrofit and real-time tracking of HVAC system are also necessary to maintain HVAC systems at their optimal performance over time. An important step towards reducing operational costs as well as meeting sustainability targets is upgrading outdated HVAC infrastructure with future generation of energy efficient components.

By using these strategies together, the HVAC industry can have a huge impact on energy efficiency and bring down carbon footprint and efforts towards global sustainability. In this future of the HVAC solution, smart technology integration, renewable energy adoption and a design characterized by advanced system design will be the focus of attention. Therefore, in response to continuous global push for net zero energy buildings and particularly in light of the increasing stringency of energy efficiency regulations, HVAC systems will have to evolve further. Innovation and investment in advanced energy saving technologies will maintain energy savings, and, at the same time will bring long term benefits in the environment and economics in buildings around the world that are both more comfortable and efficient.

5.4.2   The Future of HVAC in a Low-Carbon World

In this changing world with more and more climate change and environmental concern, the HVAC (Heating, Ventilation, and Air Conditioning) industry has come up with sustainable innovations. A considerable share of the global energy consumption is accounted by HVAC systems, and hence it is a key target for the transition to a low carbon future. Governmental bodies, corporations, and academics are entering the race to find technologies that cut emissions with low cost, energy efficiency, and comfort. Over the next decades, angst towards decarbonization, electrification, AI-driven optimization, and integration into a renewable energy world is about to change how the HVAC plays out(John W. Mitchell, 2013).

It is not just a transformation that concerns changing individual HVAC units, but changing entire heating and cooling ecosystems. Advanced thermal storage, AI powered microclimate control, waste heat recovery and smart grid integration will come into the future systems that will make them be a part of a holistic, interconnecting system where outcome will be maximum efficiency. In addition, due to the increasing energy efficiency of buildings through passive design, high performance insulation and climate adaptive architecture, the conventional HVAC will move towards hyper efficient, adaptive systems perfectly matching with the environment. Next section describes exciting innovations of groundbreaking proportions that will continue to define the role of the HVAC industry in the continued low carbon world.

Standing on the cusp of the sustainable future that the HVAC industry is set to usher in, various innovations and strategies have been converging around carrying out the efficiency drive, emission reduction and integrating renewable energy in heating and cooling systems.

Decarbonizing HVAC: The Shift Toward Fossil-Free Heating and Cooling

Due to the global trend toward building in a carbon neutral and sustainable manner, carbon neutral, fossil fuel-based HVAC systems are on the decline. Worldwide governments are laying down stricter energy efficiency regulations, carbon taxes and banning fossil fuel heating systems, which are forcing industries and home owners towards adopting electric heat pumps, district heating and geothermal energy options. Besides, technological advances like high efficiency electric boilers, infrared heating and low temperature hydronic systems compete with the traditional gas furnaces reducing carbon emissions and operational costs.

Additionally, district heating networks utilizing waste heat from industrial processes, data centres and renewable energy, including biomass and solar thermal energy, are coming into place in major cities for detailed decarbonization. With more and more electricity grids being made greener by solar, wind, and hydroelectric capacity, HVAC electrification will be crucial to eliminating the carbon emissions from heating and cooling.

HVAC as a Cornerstone of Sustainable Smart Cities

Future cities will be able to facilitate the achievement of general urban sustainability goals by leveraging interconnected city HVAC systems. Real time energy use of smart buildings is wire through built of integrated building automation control, demand response energy management and decentralized HVAC network. These developments should take place predominantly in heavily populated urban areas where reductions in energy waste and peak load management have large environmental and economic impacts.

The smart sensors along with the IoT on the HVAC systems will measure the occupancy, weather conditions, and the heat energy demand and control the heating and the cooling loads. District cooling systems, deep lake water cooling and solar assisted HVAC networks will also help in reducing the dependence on conventional air conditioning, increase in energy efficiency and reduce emissions in cities.

Biophilic and Passive Design: Reducing HVAC Dependency

By developing architectural design and biophilic building concepts, less need for active HVAC systems is being developed. Passive heating and cooling techniques can be integrated to buildings to allow them to inhabit comfortable indoor climates under small energy input. Structures are allowed to naturally keep cool in the summer and warm in the winter without too much mechanical intervention as can be extended as strategies such as natural ventilation, green roofs, thermal mass materials and optimized building orientation.

Figure 5.8: Biophilic Design in Sustainable Architecture.

Source: (BERO, 2024)

Nowadays, urban planners and architects are moving more and more towards the designs that encourage the use of the natural elements like the wind driven ventilation, evaporative cooling and solar shading to decrease the energy needs. If these strategies are implemented on a large scale, cities can reduce the need for artificial cooling and heating, and therefore HVAC systems will be a supplemental feature rather than a necessity.

AI-Powered Microclimate Control for Personalized Comfort

Real time, personal climate control will be a thing at a level never seen before with the use of AI driven HVAC systems. They will be zonal temperature adjusters—installing them in buildings, rather than single size fits all heating and cooling—and will adjust to the occupants' preferences, activity levels and physiological responses.

Smart sensors, biometric data and the use of machine learning algorithms will be used in these systems to create microclimates within a building which will continue to maintain a set temperature in individual rooms based on the room's usage. An example of this would be AI based predictive analytics whereby settings of HVAC systems are adjusted before people have arrived, warming it up if required before plus saving energy as no heating or cooling is needed. In addition, these intelligent systems will also find means to learn from occupant's habit as well as from external environmental factors to optimize their operation for optimal efficiency and comfort.

The Next Generation of Thermal Energy Storage

As we move towards an increasingly large number of renewable energy source applications, HVAC systems will need to take into account solutions to address the intermittency of these intermittent sources. These systems store excess heating or cooling energy for later use, hence lessening the need from the electricity grid during peak hours. Ice storage cooling, phase change materials (PCMs), and molten salt thermal batteries will help build energy demand for buildings push off peak time costs, making the grid more stable and cut more carbon.

In addition, underground thermal storage and seasonal heat storage will enable buildings to store and reuse energy for long periods of times, which is the hallmark of sustainable building. In future, buildings will be more self-sufficient with less dependence on outside sources of transportation or the relaxed fluctuation of energy, by integrating HVAC with smart grids and renewable energy storage.

Waste Heat Recovery and Circular Energy in HVAC

Traditional HVAC systems are extremely energy wasted. In the future, these technologies will be used to repurpose excess heat from industrial processes, data centers, refrigeration units, among others, to power space heating or electricity generation.

Figure 5.9: A regenerative thermal oxidizer (RTO) is an example of a waste heat recovery unit that utilizes a regenerative process.

Source: (WIKIPEDIA, 2021)

Buildings will be able to capture and reuse low grade waste heat, and convert what was a lost resource into an energy asset through new developments in heat to power conversion, adsorption chillers, and thermoelectric energy harvesting. These innovations are consistent with how circular energy systems run their processes; the waste from one use is the input for another, creating an enormously improved energy efficiency.

Water-Efficient HVAC: The Intersection of Energy and Water Conservation

Lots of water is used in cooling towers, evaporative cooling systems and humidifiers. Water-efficient cooling technologies that minimize water waste while maintaining a thermal performance of the future HVAC system will be integrated.

Atmospheric water harvesting, hybrid cooling systems, and air to water heat exchangers will reduce the strain on municipal water supplies, reducing the amount of stress on municipal water supplies for HVAC operations. In addition, advanced desiccant cooling systems will serve as an alternative to conventional vapor compression air conditioning with a much lower water demand while achieving efficient dehumidification and cooling.

Nanotechnology and Next-Gen Insulation Materials

There is no denying that material science has a lot to do with improving HVAC efficiency. Revolutionizing thermal management in building, aerogels, nanocoatings and super insulating vacuum panel technology are now available.

These advanced materials help make HVAC systems work less but at the same time keep the inside of the house at optimal temperature. Also, windows with electrochromic coatings can be made smart and dynamically adjust transparency for heat gain and loss, further reducing the synthetic heating and cooling.

5.4.3   Next Steps for Engineers and Facility Managers

In this spirit of the future, HVAC engineers and facility managers will maintain their active role in ingrate cutting edge technologies, consolidate existing systems, and align with the global decarbonization objectives. To transition to smart, ecofriendly, and high-performance HVAC systems, the recent regulatory changes and industry best practices, the following steps will help the professionals in the field in the effective conversion.

- **Install Smart and Adaptive HVAC Technologies:** Engineers and that manages the facilities need to adopt the use of AI driven automation, IoT connected sensors, and machine learning of transformation of heating, cooling and ventilation dynamically. By using these technologies, real time monitoring, preditional maintenance and self-adjusting allows such systems to react to occuance level and climate conditions thus maximizing energy efficiency while minimizing energy wastage.

- **Incorporate HVAC with Renewable Energy Sources:** To minimize carbon footprint, HVAC systems should be developed with incorporating Solar panels, Wind Power and Geothermal heat pumps. To enhance reliability and sustainability and reduce dependence on non-renewable energy, facility managers should explore options for thermal energy storage, hybrid heat pumps, district heating and cooling networks etc.

- Continuous data driven energy optimization should be adopted by the engineers, big data analytics, digital twins, and energy modeling software should be kept to continuously track performance metrics and redundant effects. They can fine tune HVAC operations so as to not use more resources than necessary and save costs, as well as reduce system longevity. At the same time, the remote diagnostics and optimization strategies are facilitated by cloud-based energy management platforms while buildings can also pay attention building on cloud-based energy management platforms for remote diagnostics and optimization strategies.

- Thermal performance of a building directly affects the HVAC efficiency. Engineers must work with architects and building owner to improve the insulation, enhance windows with high performance glazing and incorporate passive cool and heating techniques. These measures also keep HVAC systems in good shape by reducing the load of them during the operating season and thus saving on the operational costs and better indoor comfort.

- **Be up to date with green building standards:** The engineers and facility managers need to stay updated with the industry requirements, environmental policies and green building certifications such as LEED (Leadership in Energy and Environmental Design), WELL Building Standard, BREEAM (Building Research Establishment Environmental Assessment Method) and Energy Star. These standards not only make buildings more sustainable but also better for the market value of buildings and save energy in the long run.

- **AI Driven Automation, IoT Connected Sensors and Algorithms:** Engineers and Facility Management must utilize AI Driven Automation, IoT Connected Sensors and Algorithms in order to perform heat, cool and ventilation intelligently. These technologies allow real time monitoring, predictive maintenance, self-adjust feature that responds occupancy level, climate condition, and maintain peak energy efficiency as less energy waste as possible.

- **Carbon footprint bridging HVAC and Renewable Energy Sources:** By creating a hospitality equipment and HVAC 'synergy' with solar panels, wind power and geothermal heat pumps, the carbon footprint can be reduced. Facility managers should begin to explore thermal energy storage, hybrid heat pumps, district heating and cooling networks, etc. that may improve reliability and sustainability enough to reduce and one day eliminate natural gas.

- All of this should be done by the deployment of big data analytics, digital twins, and energy modeling software to monitor performance metrics in real time and then detect inefficiency. They will look at real time energy consumption patterns and tune HVAC operations up and save money and also extend system life. Remote diagnostics and optimization strategies are added to the list as they can be enabled through cloud-based energy management platforms that offer remote building management.

- For example, the HVAC efficiency will be directly affected by a building's thermal performance. Passive cooling and heating techniques should be used along with improvements in insulation and high-performance glazing in windows and engineers should be working with architects and building owners to achieve these goals. It enables to reduce the load on HVAC systems and reduce the operational costs and improve the indoor comfort.

- **Keep abreast of green building standards:** Engineers and facility managers ought to be abreast with the industry standards, environmental policies and green building certifications such as Leadership in Energy and Environmental Design (LEED), WELL Building Standard, Building Research Establishment Environmental Assessment Method (BREEAM) and Energy Star. You may consider that these standards don't only make the buildings more sustainable but also save energy in the long run, and better for the market value of buildings.

5.5   Chapter Synthesis

In this chapter we discuss the emerging technologies in the field of HVAC energy efficiency such as digital twin, AI, and blockchain that improves the system optimization, predictive maintenance and the energy management. For personal comfort and efficient use of energy, advanced climate control, smart thermostats and zoning technology are used together. In smart cities, real time data is then also used by AI-driven HVAC systems to raise energy savings efficiency. Blockchain is also used to secure, transparent and peer to peer transaction in the energy systems. Sustainability is driven by innovations of variable speed compressors, high-efficiency heat exchanger, air purification technologies, as well as adaptive controls. This integration of the AI, IoT, and cloud-based platforms improves the performance and reduces the down time as well as improving the predictive maintenance of machines. The activities considered as sustainable HVAC strategies are high efficiency equipment, renewable energy integration, passive design strategies and advanced thermal storage. Adopting the electric heat pump, the district heating, the smart grid in order to eliminate the fossil fuel is the scene of the adoption of carbon neutral buildings. In future, we will develop towards AI powered micro climate control, waste heat recovery, water efficient HVAC technologies and next generation insulation materials for extending energy efficiency and sustainability of smart buildings and cities.

Multiple Choice Questions (MCQs)

1. **What is a key advantage of using digital twins in HVAC systems?**

   a. They eliminate the need for maintenance
   b. They allow engineers to optimize system design before installation
   c. They replace physical HVAC systems with virtual ones
   d. They increase energy consumption

2. **How do AI-driven HVAC systems improve energy efficiency in smart cities?**

   a. By controlling temperature manually in different buildings
   b. By utilizing real-time data, weather predictions, and occupancy patterns
   c. By reducing the need for insulation in buildings
   d. By eliminating the need for human supervision

3. **What is the primary role of blockchain in HVAC energy management?**

   a. Increasing the physical size of HVAC units
   b. Ensuring secure, transparent, and decentralized energy transactions
   c. Replacing AI and IoT in HVAC systems
   d. Reducing humidity levels in smart buildings

4. **Which of the following is a best practice for engineers and facility managers in HVAC energy efficiency?**

   a. Ignoring predictive maintenance alerts
   b. Using outdated refrigerants in HVAC systems
   c. Implementing data-driven monitoring and optimization strategies
   d. Avoiding smart controls to reduce system complexity

5. **What is a key strategy for HVAC systems in a low-carbon world?**

   a. Increasing reliance on fossil fuel-based heating
   b. Integrating renewable energy sources like solar and geothermal heating
   c. Eliminating ventilation systems in buildings
   d. Reducing AI integration in HVAC systems

6. **How can engineers and facility managers prepare for the future of HVAC energy efficiency?**

   a. Avoiding new technology implementation to reduce costs
   b. Adopting AI-driven automation and IoT sensors for better system control
   c. Increasing energy consumption for comfort optimization
   d. Eliminating zoning technology in HVAC systems

7. **What role does virtual commissioning play in HVAC energy efficiency?**

   a. It allows for HVAC system testing and optimization before physical installation
   b. It replaces the need for HVAC maintenance teams
   c. It increases the energy demand of HVAC systems
   d. It eliminates the need for AI in HVAC operations

8. **Which of the following technologies can help reduce energy waste in HVAC systems?**

   a. Smart thermostats and adaptive climate controls
   b. Fixed-speed compressors and outdated insulation
   c. Manual temperature regulation in buildings
   d. Increasing HVAC unit size without efficiency considerations

9. **Why is HVAC transitioning towards district heating and geothermal energy solutions?**

    a. To increase dependence on fossil fuels
    b. To improve energy efficiency and reduce carbon emissions
    c. To make HVAC systems more expensive to maintain
    d. To eliminate the use of AI in smart buildings

Answers

1	2	3	4	5	6	7	8	9	10
b	d	b	c	b	b	a	a	b	b

BIBLIOGRAPHY

Abiz. (2024). *How Building Automation Systems (BAS) Work*. Abiz.

Aemaco. (2024). *The Role of HVAC Energy Management Systems in Commercial Buildings*.

Aireserv. (2024). *The Basics of HVAC Systems: A Comprehensive Guide for Beginners*. Aireserv.

Al Momani, D., Al Turk, Y., Abuashour, M. I., Khalid, H. M., Muyeen, S. M., Sweidan, T. O., Said, Z., & Hasanuzzaman, M. (2023). Energy saving potential analysis applying factory scale energy audit – A case study of food production. *Heliyon*, *9*(3), e14216. https://doi.org/10.1016/j.heliyon.2023.e14216

Asgari, S., Gupta, R., Puri, I. K., & Zheng, R. (2021). A data-driven approach to simultaneous fault detection and diagnosis in data centers. *Applied Soft Computing*, *110*, 107638. https://doi.org/https://doi.org/10.1016/j.asoc.2021.107638

ASHRAE. (2020). Variable Refrigerant Flow. In *ASHRAE handbook : heating, ventilating, and air-conditioning systems and equipment*.

BERO, A. (2024). *What is Biophilic Design? Trends & Examples in 2024*. Gb&d.

Bhatti, S. S., Kumar, A., R, R., & Singh, R. (2024). Environment-Friendly Refrigerants for Sustainable Refrigeration and Air Conditioning: A Review. *Current World Environment*, *18*, 933–947. https://doi.org/10.12944/CWE.18.3.03

Burdick, A. (2012). Strategy guideline: Accurate heating and cooling load calculations. In *Guidelines for Improved Duct Design and HVAC Systems in the Home*.

carrier. (2023). *DUCTLESS MINI SPLIT ACs AND HEAT PUMPS*. Carrier.

Click maint. (2023). Predictive Maintenance Improving Equipment Reliability. In *Collection of scientific works of Odesa Military Academy* (Issue 20, pp. 51–55). https://doi.org/10.37129/2313-7509.2023.20.51-55

Compny, U. (2024). *What is a High Efficiency Gas Boiler?* Usboiler Company.

Connor Holbert. (2025). *Maximizing HVAC efficiency: A guide for sustainable building management.* Connor Holbert.

Continal. (2020). *The benefits of underfloor heating.* Continal.

CoolAutomation. (2019). *VRV or VRF?* Coolautomation.

dailycivil. (2024). *Thermal Insulation in Buildings | Types and Materials.* Dailycivil.

dandb. (2023). Eco-Friendly HVAC Systems: Your Green Heating & Cooling Guide. *Dandbclimatecare.*

Discoveries, E. (2024). *building to face south (in the Northern Hemisphere) can maximize solar heat during the winter, reducing heating needs. Cooling: In warmer climates, orienting the building to minimize exposure to the sun (e.g., east or west-facing facades) can reduce cooli.* Engineering Discoveries.

ECBC. (2001). 3 . ENERGY MANAGEMENT AND AUDIT Energy Audit : Types And Methodology. *Energy*, 54–78.

EDF. (2024). *Types of renewable energy.* Edfenergy.

Editorial, T. (2023). *Harness solar power for energy-efficient HVAC systems.* Thermal Control Business Update.

EDS. (2024). *How an HVAC Load Calculator Optimizes System Performance.* EDS.

efficiencymaine. (2024). *Boilers and Furnaces.* Efficiencymaine.

Ehrlich, P. (2018). Building automation systems innovation. *Engineered Systems, 35*(11), 22–27.

Elion. (2024). *Case Studies_ Successful Energy Audits in Indian Industries.*

energysaver. (2023). *Geothermal Heat Pumps.* Energy.

EPA. (2024). *Introduction to Indoor Air Quality.* EPA.

Flickr. (2022). *Boiler and Hot Water Heater.* Flickr.

Francis, T. &. (2004). *Sustainable Architectures Cultures and Natures in Europe and North America.*

Goulden, M., & Spence, A. (2015). Caught in the middle: The role of the Facilities Manager in organisational energy use. *Energy Policy, 85,* 280–287. https://doi.org/10.1016/j.enpol.2015.06.014

Gulvin, L. (2023). *Preventive maintenance vs. predictive maintenance.*

henick. (2023). *The Importance of HVAC Energy Efficiency.* Henick-Lane.

Hitachi. (2023). *NP Series*. Hitachi.

Hosseini Gourabpasi, A., & Nik-Bakht, M. (2024). BIM-based automated fault detection and diagnostics of HVAC systems in commercial buildings. *Journal of Building Engineering, 87*, 109022. https://doi.org/https://doi.org/10.1016/j.jobe.2024.109022

HVAC Automation. (2024). *The Role of Artificial Intelligence in Modern HVAC Systems*. Mgcs.

Ian Dempster. (2018). *Barriers to HVAC System Optimization and How to Overcome Them*. Coolingbestpractices.

Indiamart. (2024). *Hvac Efficient Energy Management Services*.

Innovations, S. S. (2021). *Home Automation System*. India Mart.

Instructor, M. A. (2022). *How a Variable Air Volume VAV System Works*. Mepacademy.

iqsdirectory. (2023). *Chillers and Chiller Units: Types, Uses and Descriptions*. Iqsdirectory.

Jane Marsh. (2024). *How Smart HVAC Helps Smart Cities Be Sustainable*. Greencitytimes.

Jasim, A. M., Jasim, B. H., Neagu, B. C., & Alhasnawi, B. N. (2023). Efficient Optimization Algorithm-Based Demand-Side Management Program for Smart Grid Residential Load. *Axioms, 12*(1). https://doi.org/10.3390/axioms12010033

Jaysawal, R. K., Chakraborty, S., Elangovan, D., & Padmanaban, S. (2022). Concept of net zero energy buildings (NZEB) - A literature review. *Cleaner Engineering and Technology, 11*, 100582. https://doi.org/https://doi.org/10.1016/j.clet.2022.100582

John W. Mitchell, J. E. B. (2013). *Principles of Heating, Ventilation, and Air Conditioning in Buildings*.

Kalbasi, R., Samali, B., & Afrand, M. (2023). Taking benefits of using PCMs in buildings to meet energy efficiency criteria in net zero by 2050. *Chemosphere, 311*, 137100. https://doi.org/10.1016/j.chemosphere.2022.137100

Katlyn Avery. (2024). *Discover the three types of thermal energy storage systems*. Araner.

KG. (2023). How to Perform Load Calculations on Column, Beam, Wall, and Slab: A Comprehensive Guide. *Linkedin*.

Kinetik. (2011). The ultimate guide to Green Formulating. In *New Scientist* (Vol. 212, pp. 42–47).

Kumar Singh, A., Kumar, V. R. P., Dehdasht, G., Mohandes, S. R., Manu, P., & Pour Rahimian, F. (2023). Investigating the barriers to the adoption of blockchain technology in sustainable construction projects. *Journal of Cleaner Production, 403*, 136840. https://doi.org/https://doi.org/10.1016/j.jclepro.2023.136840

Kumari, A., Kakkar, R., Gupta, R., Agrawal, S., Tanwar, S., Alqahtani, F., Tolba, A., Raboaca, M. S., & Manea, D. L. (2023). Blockchain-Driven Real-Time Incentive Approach for Energy Management System. *Mathematics, 11*(4), 928. https://doi.org/10.3390/math11040928

Lubach, D. (2024). *HVAC Trends: What Facility Managers Need to Know.* Facilities Net.

Mannan Rana. (2024). Design and Optimization of Energy-Efficient HVAC Systems for Smart Buildings. *International Journal for Research Publication and Seminar, 15*(4), 50–59. https://doi.org/10.36676/jrps.v15.i4.5

Manvendra Kunwar. (2024). *IoT in HVAC: Transforming the Way We Manage Heating, Ventilation, and Air Conditioning.* Hashstudioz.

Mărgulescu, S., Mărgulescu, E., Kardontchik, J. E., Ave, S. B., Khan, A., Nicholson, J., Mellor, S., Jackson, D., Ladha, K., Ladha, C., Hand, J., Clarke, J., Olivier, P., Plötz, T., Yang, X., Zhong, K., Zhu, H., Kang, Y., Mishra, A. K., … Haghighat, F. (2014). Characteristics of buoyant flow from open windows in naturally ventilated rooms. *Energy and Buildings.*

Mike Haine. (2023). *How Geothermal Heat Pumps Work.* Acdirect.

Ministry of Energy Transition. (2021). *Strategy for Energy Retrofitting of National Building Stock: Italy. March*, 1–66.

Mishra, P., & Singh, G. (2023). Energy Management Systems in Sustainable Smart Cities Based on the Internet of Energy: A Technical Review. *Energies, 16*(19). https://doi.org/10.3390/en16196903

Nafeez khan. (2023). *Alternative Sources of Energy.* Vedantu.

Nwagu, C. N., Ujah, C. O., Kallon, D. V. V, & Aigbodion, V. S. (2025). Integrating solar and wind energy into the electricity grid for improved power accessibility. *Unconventional Resources, 5*, 100129. https://doi.org/https://doi.org/10.1016/j.uncres.2024.100129

O'Keefe, P., O'Brien, G., & Pearsall, N. (2010). The future of energy use. In *The Future of Energy Use*. https://doi.org/10.4324/9781849774819

Obinna Iwuanyanwu, Ifechukwu Gil-Ozoudeh, Azubuike Chukwudi Okwandu, & Chidiebere Somadina Ike. (2024). Retrofitting existing buildings for sustainability: Challenges and innovations. *Engineering Science & Technology Journal*, *5*(8), 2616–2631. https://doi.org/10.51594/estj.v5i8.1515

Paddeco. (2021). Introduction to an Air Handling Unit (AHU). *SQA Approved.*

Palombo, A. (2002). Comparison between total and sensible energy recovery HVAC systems for the Italian climate. *International Journal of Ambient Energy*, *23*(2), 79–96. https://doi.org/10.1080/01430750.2002.9674874

Patel, N., & Buddhi, D. (2022). *Analysis on Energy-Efficient HVAC System for Buildings* (pp. 213–218). https://doi.org/10.1007/978-981-16-3132-0_21

Pathshala. (2024). *Module 14 : Performance Monitoring.*

Pro, C. (2024). *Emerging HVAC Technologies: Revolutionizing Comfort in Modern Living.* Cool Pro.

Reddy, T. A., Kreider, J. F., Curtiss, P. S., & Rabl, A. (2016). *Heating and Cooling of Buildings.* CRC Press. https://doi.org/10.1201/9781315374567

REvolution, G. (2020). *Solar Thermal HVAC.* GReen Revolution.

Richert, B. (2022). *How To Calculate The Ventilation Rate For A Confined Space.* Sciencing.

Rob Ezold. (2024). *Thermal Energy Storage for Chilled Water Systems.* Vertexeng.

RSI. (2024). *What Is Geothermal HVAC and How Does It Work?* RSI.

Ryalat, M., Franco, E., Elmoaqet, H., Almtireen, N., & Al-Refai, G. (2024). The Integration of Advanced Mechatronic Systems into Industry 4.0 for Smart Manufacturing. *Sustainability (Switzerland)*, *16*(19), 1–39. https://doi.org/10.3390/su16198504

Sandia. (2023). *Thermal Energy Storage.* Sandia National Laboratories.

SAP. (2024). *What is predictive maintenance_ A complete overview.*

SBA. (2024). *(12) HVAC Energy Management_ Strategies for Optimizing HVAC Systems.*

Sean McCleland. (2024). *The Role of Renewable Energy in HVAC Systems.* Airoptions.

Service, 1ˢᵗ Choice pro. (2024). *A Comprehensive Guide to HVAC Energy Audits.*

Seyam, S. (2018). Types of HVAC Systems. In *HVAC System.* InTech. https://doi.org/10.5772/intechopen.78942

Sharma, P., Reddy Salkuti, S., & Kim, S.-C. (2021). Energy audit: types, scope, methodology and report structure. *Indonesian Journal of Electrical Engineering and Computer Science, 22*(1), 45. https://doi.org/10.11591/ijeecs.v22.i1.pp45-52

Sharma, R. K., Kumar, A., & Rakshit, D. (2024). A phase change material (PCM) based novel retrofitting approach in the air conditioning system to reduce building energy demand. *Applied Thermal Engineering, 238,* 121872. https://doi.org/10.1016/j.applthermaleng.2023.121872

Sharma, V., & Mistry, V. (2023). *Automated Fault Detection and Diagnostics in HVAC Systems.* https://doi.org/10.5281/zenodo.11079964

Shyam Varan Nath, Pieter van Schalkwyk, D. I. (2021). *Building Industrial Digital Twins Design, Develop, and Deploy Digital Twin Solutions for Real-world Industries Using Azure Digital Twins.*

Sketchbubble. (2024). *Preventive Maintenance vs Predictive Maintenance vs Proactive Maintenance.*

slipstreaminc. (2023). *Variable Refrigerant Flow (VRF).* Slipstreaminc.

smarterhouse. (2024). *Ventilation and Air Distribution.* Smarterhouse.

space. (2023). *Sustainable Refrigeration.* Space.

sprsunheatpump. (2024). *Types of Heating Systems You Should Know.* Sprsunheatpump. https://doi.org/https://sprsunheatpump.com/types-of-heating-systems.html

Stanford III, H. W. (2016). HVAC Water Chillers and Cooling Towers. In *HVAC Water Chillers and Cooling Towers.* https://doi.org/10.1201/b11510

Steinbrecher, R., & Schmidt, R. (2011). Data Center Environments: ASHRAE's Evolving Thermal Guidelines. *Ashrae Journal, 53,* 42–49.

Technical Editor. (2024). *Orientation of building.* Polytechnichub.

TERI. (2019). *Existing Commercial Building Retrofit Guidance.* 60pp.

Tosin Michael Olatunde, Azubuike Chukwudi Okwandu, Dorcas Oluwajuwonlo Akande, & Zamathula Queen Sikhakhane. (2024). Review of energy-efficient HVAC technologies for sustainable buildings. *International*

Journal of Science and Technology Research Archive, 6(2), 012–020. https://doi.org/10.53771/ijstra.2024.6.2.0039

Trance. (2023). *Hybrid Heat Systems: How Do They Work?* Trane.

Travis Baugh. (2024). *How High Efficiency Furnaces Work.* Bryan.

Trout, J. E., & Betts, C. P. (1988). Security Engineering. *Military Engineer*, *80*(522), 368–371. https://doi.org/10.4018/978-1-59140-911-3.ch011

tutorialspoint. (2024). *Demand Factor, Load Factor, and Diversity Factor.* Tutorialspoint.

Ucar, A., Karakose, M., & Kırımça, N. (2024). *applied sciences Artificial Intelligence for Predictive Maintenance Applications :*

velosiot. (2024). *Smart HVAC: How IoT Transforms Smart HVAC Systems.* Velosiot.

Vikram Murthy. (2024). *AI and HVAC – An Introduction to What's in Store.* International Academic Journal of Human Resource and Business Administration.

VR, A. (2023). *The Benefits and Challenges of Solar-Powered Refrigeration and Air Conditioning.* ARkA.

Wan, H., Cao, T., Hwang, Y., & Oh, S. (2020). A review of recent advancements of variable refrigerant flow air-conditioning systems. In *Applied Thermal Engineering.* https://doi.org/10.1016/j.applthermaleng.2019.114893

wbdg. (2023). *Geothermal Heat Pumps.* Wbdg.

WIKIPEDIA. (2021). *Waste heat recovery unit.*

About the Author

Jaymin Shah is a dedicated and results-driven Energy Engineer with extensive experience in energy management, auditing, and sustainability. With a strong foundation in mechanical engineering and advanced expertise in energy systems, he has made significant contributions to improving energy efficiency across various sectors, including residential, commercial, and institutional buildings.

Jaymin holds an Executive MBA from the University of the Cumberlands and a Master of Science in Energy Management from the New York Institute of Technology. His academic background has provided him with a strategic approach to energy management, combining technical knowledge with business acumen to drive impactful energy solutions.

With a career spanning several years, Jaymin has led and conducted over 150 ASHRAE Level I & II energy audits, achieving notable energy savings and cost reductions. He has played a crucial role in Local Law 87 compliance audits, optimizing energy performance for high-rise residential and commercial properties. His expertise extends to commissioning for new and existing buildings, ensuring optimal system functionality and efficiency.

Throughout his career, Jaymin has worked with industry leaders such as FST Technical Services, ICF Consulting, and Greater Bergen Community Action, where he has been instrumental in identifying and implementing energy conservation measures (ECMs) that enhance building performance. His work has resulted in significant energy reductions, with projects achieving up to 25% savings in energy usage.

As a Certified Energy Manager (CEM), Project Management Professional (PMP), and LEED Green Associate, Jaymin is committed to sustainability and the advancement of renewable energy technologies. He has experience in energy benchmarking, utility tariff analysis, HVAC optimization, and financial feasibility studies for energy projects. His ability to manage multiple projects while ensuring high client satisfaction has been a key factor in his success.

Beyond his technical expertise, Jaymin is a thought leader and advocate for energy efficiency and green building practices. He actively participates in industry conferences, collaborates with stakeholders to promote sustainability initiatives, and stays updated on evolving energy policies and technologies.

Jaymin's passion for energy efficiency, combined with his strategic mindset and hands-on expertise, makes him a valuable asset in the field of energy management. His journey is a testament to his dedication to creating sustainable, cost-effective, and high-performing energy solutions for a better future.